The Netweaver's Sourcebook

A Guide to Micro Networking and Communications

DEAN GENGLE

 Addison-Wesley Publishing Company

Reading, Massachusetts . Menlo Park, California
London . Amsterdam . Don Mills, Ontario . Sydney

To the future:
Amy, Guy, Jennifer, Julie, and Justin

Many of the designations used by manufacturers and sellers to distinguish their products are claimed as trademarks. Where those designations appear in this book and Addison-Wesley was aware of a trademark claim, the designations have been printed in initial caps (e.g., VisiCalc) or all caps (e.g., UNIX).

Library of Congress Cataloging in Publication Data

Gengle, Dean.
  The netweaver's sourcebook.

  Bibliography: p.
  Includes index.
  1. Data transmission systems.   I.   Title.
TK 5105.G46   1984   384   84–6212
ISBN 0–201–05208–3

ISBN 0–201–05208–3

ABCDEFGHIJ–HA–8987654

Preface

Recent years have witnessed explosion on top of explosion in the numbers and kinds of technology-related social changes happening in all segments of our society. The advent of low-cost, mass-produced microprocessors is part of the "first explosion." Whole industries have sprung up around applications for microcomputers based on our ability to mass-produce these complex circuits on a chip. The initial applications, for the most part, have merely been new ways to make micros do our old tasks. Even so, the changes that these applications have brought about are monumental: personal, social, global transformations. This has led to a condition that one writer has deemed "running wild." It is also referred to as the "home computer revolution," although relatively few homes actually have revolted. Mostly offices are revolting, have revolted.

In any case, if you are one of the five million or so people who already own or have access to a personal micro, or are one of the million or more who will get one this coming year, then you are already tuned in to this first wave of change.

As this first transformation begins to peak, a second set of changes is thereby rendered inevitable, so look out. Subsumed under the term "telecommunications," this second explosion is a result of the marriage of home micros, television, and increased capabilities in the telephone system, especially those brought on by satellite communications. It is also spurred on, in the United States, at least, by sweeping changes in the regulations heretofore governing the communications industries. Entrenched telephone (AT&T) and broadcast (NBC, CBS, ABC) and cable (HBO, Westinghouse, RCA) companies are being challenged by a host of younger, technologically sophisticated newcomers in the marketplace. As they say, all hell is breaking loose down here. More changes in regulation are on their way, promising to make market conditions even hotter, not to mention the political debates in the halls of our great Evolutionary Drag: Congress.

Books on the subjects of teletext, personal computer communications, and the so-called home micro revolution have been rushed into print in order to take advantage of increasing interest in and need for adequate guides to these new sociotechnical developments. Alas, most of them suffer from the fact that they attempt to accommodate individuals to what has already happened rather than helping them take full advantage of the new possibilities inherent in the technology. It is inherent also in the speed of these things that new possibilities are being generated every day, so people need ways of tuning in to trends rather than having to swim through a sea of detail themselves.

There are few enough guides around, to begin with. Add to that the fact that all such purported guides emphasize the technical at the expense of the individual and social dynamics involved. To emphasize the theme: we are talking about fundamental changes in the way human beings communicate, work, and play. Is that sweeping enough for all you rebels? Unsettling? Then you must continue.

This sourcebook addresses the telecommunications transformation from the point of view of a designer. I have chosen to call this hypothetical designer—you—a *netweaver*. If you find that too presumptuous, do pretend I'm speaking to someone else. You won't find extensive listings of commands and sample sessions from existing mass information utilities here. Instead, you will find a helpful companion in the journey toward creating and using your own information system, which, eventually, you can expand to include some pretty wonderful global resources. Through this sourcebook you can begin the process of understanding and gaining command of the dynamics of telecommunications. These dynamics include hardware, software, switches, connections, legal help, organizations, day-to-day maintenance of your network, databases, directories, local area networks, global networks, telephones, videodisks, downlink satellites, microwaves, protocols, and, most important, how to find out more about and keep up with changes as they occur in these diverse and interesting matters.

To be a netweaver—to create your own communications networks based on your own needs and goals—you do not need to be a computer programmer or hardware genius. You do not need to know any computer languages. Although there are some "tricks of the trade," you'll find many of those in this book and/or discover/create most of them as you proceed. If you haven't yet gotten your own micro, this book should be all the argument you will need to justify its purchase, in case you still need such an argument.

A fundamental premise of this book is that our society has one foot in the Information Age and the other in the fifteenth century. This creates collective symptoms of what social critics, mystics, philosophers, novelists, futurists, and academics have been telling the human race for several years now: that our scientific and technological powers have far outstripped our moral and intuitive/spiritual and ethical capacities. If the latter do not develop, we have been warned over and over again, we will destroy ourselves with the former.

Enhanced communications ability is a necessary, built-in precondition and major feature of the Information Age. Enhanced ability to personally come to grips with a diversity of necessary information is a good reason for investing more dollars and energy in your own information system.

If you already have a micro, know a computer language or two, or have some telecommunications experience, this book will help you bring it all together so you can build on that foundation. Finally, if you are interested in helping others create their own information systems and networks, this book contains information that is not readily available anywhere else in a single source.

ACKNOWLEDGMENTS

This book has been three real-time years and several more person-years in the making. Much of the information it contains has come out of the CommuniTree Network, a system of small-scale microcomputer communications centers throughout the country. I am, first and foremost, indebted to the operators and users of these systems—far too many to list here—for the learning and experience they have shared.

Much credit is due Dr. Scot Kamins and my Hawthorne colleagues in study whose encouragement, advice, and support helped to launch this book out of the infovoid and into the real world.

Steven Smith, John James, and John Mellen have my perpetual thanks for being visionary enough to agree that this book needed to be written, and for their invaluable ongoing feedback.

For their help on the technical front lines, gold cable awards to Wade Myers and Bob Kouré. Thank you.

Where would writers be without good editors? Nowhere, that's where. So I also gratefully acknowledge Ted Buswick, my editor and guide at Addison-Wesley.

The data base on which this book rests is huge and would not be as comprehensive as it is without the constant attention and organizing ability of research colleague Lynne Ashdown.

While the input of many has gone into this book, responsibility for errors of omission, commission, and emission should be pinned entirely on the author.

Weave on!

D.G.
San Francisco, 1984

Contents

Introduction

> Therefore, those who dismiss the issues involved in computing as mere philosophy do so at their own risk.
>
> —Covvey and McAlister, *Computer Choices*

WHY YOU NEED THIS BOOK

This sourcebook is a cybernetic [1.6.4] manual that will help you navigate your immediate future. You need this book because it can save you time in the months to come. It can also save you dollars and, possibly, suggest to you new ways to work and make yourself and your company more productive. It can save you from being "the last to know." In our society's continuing transformation [1.3] from industry to information [1.6.5], being last to find out, being the last to get vital information, can be costly indeed.

You need this book if you just want to find out how to get your micro at home to talk to your office computer [4.0]. If you just want your micro system to work harder for you, instead of the other way around, then this sourcebook is for you.

HOW THIS BOOK IS ORGANIZED

The Netweaver's Sourcebook is organized around four chapters:

1. *Netweaving Development Tools:* in which netweavers learn about the languages and general concerns of information and networks [1.1, 1.9]. When dealing with a socially and culturally dynamic topic such as netweaving, there are no simple, direct "how-to" instructions or procedures. There are only rules of thumb and new solutions. To fully appreciate all the options and implications open to the netweaver, a common base of general theory and terminology is necessary, as well as some appreciation of the historical background of these theoretical developments.

2. *Personal Systems:* in which netweavers are centered in the most basic information system—the self and its brain [2.1]. It is this component of information systems—you—that makes them "living systems" [1.6]. This important element is often ignored in favor of overemphasis on nonliving human artifacts: hardware.

3. *Social Systems:* in which netweavers learn some of the basics of human organization, starting with organizing oneself [3.3] and from there working with small groups [3.4] and organizations [3.9]. Getting things done in time and space, using material objects to accomplish an end, requires coordination and control. Computers and networks can assist managers in guiding processes into purposes while still maintaining and encouraging the expansion of individual freedoms. These freedoms are crucial elements in the evolving global information economy and are central to netweaving.

4. *Building and Weaving:* in which the netweavers learn the basics of the hardware, software [4.14], and physical connections [4.13] they'll be dealing with over the course of building new information systems and weaving new networks.

Each chapter is broken into numbered sections, each with a descriptive word or keyphrase at the beginning. These keyphrases let you find your way around quickly.

Four appendixes correspond to the four chapters of the text. The appendixes contain annotated references for the sections and subsections. Appendix IV contains volatile information, more likely to change or be outdated quickly, about representative hardware, software, and related technical resources. Appendix IV is *not* intended to be a "comprehensive directory." Rather, items have been selected for inclusion based on qualitative criteria:

- They are excellent examples of their kind.
- They are essential for the netweaver to know about.
- They are representative of a future trend or possibility.
- They are just plain useful to know about.

HOW TO USE THIS BOOK

Begin with those items that interest you most and proceed from there. The book can be used as a map or operating manual might be used. However, the first chapter, Netweaving Development Tools, presents concepts used throughout the remaining sections, so you might want to begin your journey there. Cross-references are in square brackets, e.g., [1.0]. These references will point you back to items or terms you may

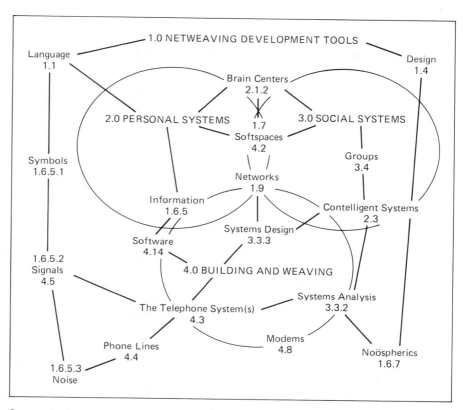

Semantic Network—*The Netweaver's Sourcebook.* Another kind of network, this semantic map shows one possible configuration of the contents of this book. The lines connecting the ideas are lines of association. The outside edges show *contextual* ideas that form the background on which the major sections of the book "float." Not all the sections are listed on this map. Other associations, of course, could also be made. The human brain is capable of storing an enormous number of such idea maps in its neural networks.

have missed, and sometimes forward to related useful items. If you already know the term or just want to speed-read, just skip over the bracketed cross-references.

Some subsection descriptive keywords may be used more than once. These are connecting ideas that serve as major and minor themes throughout the sections. The material covered is usually slightly different or viewed from a slightly changed perspective.

This book can be used as a *planning tool.* If you are thinking about installing or enhancing an existing personal information system, including but not limited to a personal micro, the various section checklists, coupled with the Building and Weaving chapter, can assist you in planning your approach. If you already have a micro but haven't yet used it in telecommunications, then the Building and Weaving section and appendixes will help you enter this exciting new universe. If you are already "into" telecommunications with your micro, then *The Netweaver's Sourcebook* will be valuable in suggesting new things to do, streamlining your existing uses, and helping to make these new possibilities available to others in your family, school, or business.

If you are a researcher, or someone who uses the microcomputer and networks as mere tools toward your own professional or personal ends, this book can connect your specialty and the infrastructure that can make the practice of your science, skill, art, or craft a much more enjoyable and productive activity.

If you plan a career in telecommunications or some other microcomputer-related business, then this sourcebook can serve as an essential connecting document in your softspace [1.7]. It will give you an overview and a set of concepts that will help you organize the constant flood of new information relating to microcomputers, networks and telecommunications. It will only give you a start. The rest is up to you.

It is my sincere hope that we will outgrow any need there might be for full-time, "professional" netweavers or networkers in the future. Rather, these activities, combined with your own interests, aptitudes, and skills, should be seen as appropriate technology for application as the need arises. However, if you are looking for new career or business directions, the netweaver's arts may just suggest some new paths to take.

MAJOR PREMISES

In preparing this sourcebook, several assumptions and ground rules had to be established. Netweaving, networks, telecommunications, and such are parts of dynamic living systems in which every individual takes part. What you hold in your hands is a "snapshot"—a cross-section in time—that will enable you to benefit from our accumulation and synthesis of information. Making some of our starting points and ground rules explicit will help you get maximum use out of the book.

- This book was prepared for personal micro owners and those considering buying a personal micro for use at home, in a small business, or in a corporate setting. It assumes that you already have a general overview of what the micro itself is and generally how it does what it does when it does it.

- The primary uses of personal micros are (1) enhanced communications ability and (2) enhanced personal information handling ability.

- It is not only possible but imperative that individuals take charge of the new media technologies. This may and probably will require

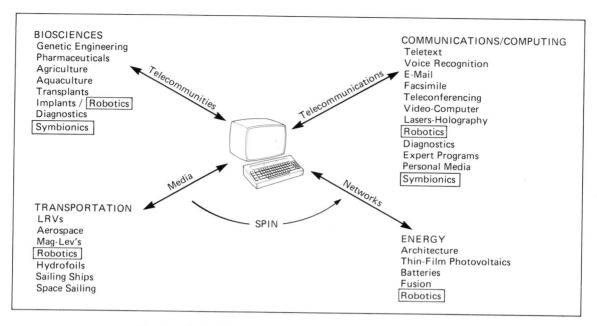

BIOSCIENCES
Genetic Engineering
Pharmaceuticals
Agriculture
Aquaculture
Transplants
Implants / Robotics
Diagnostics
Symbionics

COMMUNICATIONS/COMPUTING
Teletext
Voice Recognition
E-Mail
Facsimile
Teleconferencing
Video-Computer
Lasers-Holography
Robotics
Diagnostics
Expert Programs
Personal Media
Symbionics

Telecommunities   Telecommunications

Media   Networks

SPIN

TRANSPORTATION
LRVs
Aerospace
Mag-Lev's
Robotics
Hydrofoils
Sailing Ships
Space Sailing

ENERGY
Architecture
Thin-Film Photovoltaics
Batteries
Fusion
Robotics

The Art of the Netweaver. The four "hottest" areas of activity and need in the current phase of the Information Age: the biosciences, transportation, communications/computing, and energy. Each is interacting with and impacting on the others in highly synergetic ways. The four "linking concepts" are: networks, telecommunities, telecommunications, and media. Robotics brings together ideas and applications from all four areas, which is why it's highlighted here. SPIN stands for SPecial Interest Network, and also indicates that there's a dynamic quality to this otherwise static model. The shape of the "artpiece" in the center will vary from netweaver to netweaver.

change on the part of the person attempting to do so. Whether the arena in question is at home or at an office of one's own, or at an office in which others also work, makes little difference in terms of the actual change requirements involved.

- Making and managing your own change is made easier if you have access to certain scientific (technical), psychological, and organizational information. Technical information by itself is useful but less likely to be applied where there is resistance to change emanating from other realms of valid human experience.

- Learning is a fundamentally enjoyable human activity.

- Imagination is the primary source of wealth in the Information Age.

- The general-purpose information machine—a microcomputer—is the primary tool or "handle" with which to take charge of the new media technologies.

- Serious netweavers—that's you, remember?—will gently point out the errors in this book and will help correct them based on the following question: Does this information correspond to my current best experience in the matter? (See the Netweaver's Feedback Form at the back of the book.)

1

Netweaving Development Tools

Thought is crude, matter unimaginably subtle. Words are few and can only be arranged in certain conventionally fixed ways; the counterpoint of unique events is infinitely wide and their succession indefinitely long. That the purified language of science, or even the richer purified language of literature should ever be adequate to the givenness of the world and of our experience is, in the very nature of things, impossible. Cheerfully accepting the fact, let us advance together, people of letters and people of science, further and further into the ever-expanding regions of the unknown.

—Aldous Huxley, *Literature and Science*

Before entering any new information domain it is helpful to know how it relates to broader concerns. "Netweaving Development Tools" presents *organizing concepts* that provide netweavers with the mental scaffolding needed before they can deal successfully with the wealth of detailed information emerging hourly in the field. This framework will make it easier to construct, reorganize, and manage these details. If you wish, you can simply scan this section. As you review the ideas in it, bear in mind that one or more may be useful to you later on as you undertake to accomplish the various tasks and subtasks involved in organizing a telecommunications network. This conceptual framework will assist netweavers (those who design and build) and networkers (those who use and appreciate) alike in pursuing their respective ends. Later on, you just might replace this framework altogether in favor of one of your own.

1.1 LANGUAGE

We can all either contribute to our common store of facts and our logical use of them, or swell the falsehoods and the irrational appeal of propaganda, advertisement and demagogy. All the most important problems, the problems of religion, morals, politics, and sociology, can only be solved via the use of words. To understand the use of words properly is plainly a prior condition of solving them successfully.

—John Wilson, *Language and the Pursuit of Truth*

The human universe is cradled in language. It is a universe long since displaced from the "natural," as environmentalists might use the word. Immersed as we are in culture, forms arise in our brains and are contained in speech and shuffled around as we find words for continually new experiences. The persons and objects around us are no longer completely within the world we call natural. They are subject to the workings of the human brain first and foremost. The past can be caused to haunt the present. Gods and goddesses murmur in the trees. Druids, witches, shamans, and sorcerers still walk the earth. Nonetheless, we twine a web of sustaining mathematics around it all and pretend that we are being "realistic."

Forms arise in the brain. Sometimes the forms consist of little more than an intuition, a whisper of a hint. The forms may consist of words and sentence fragments or visual images (diagrams), or both. For the sake of convenience and memory, we may devise new words to describe new forms. Thus are neologisms born.

Neologism. (1) A newly coined word, phrase, or expression, or a new meaning for an old word. (2) The use of new words, phrases, or expressions or of new meanings for old words.

1.1.1   Jargon

The following definitions distinguish between three types of jargon:

Jargon of the First Kind. Nonsensical, incoherent, or meaningless utterance; gibberish.

Jargon of the Second Kind. A hybrid language or dialect; pidgin.

Jargon of the Third Kind. The necessarily specific language of a trade, profession, class, or fellowship.

The word "jargon" derives from Middle English and Old French with overtones of twittering, meaningless chatter. Those who use jargon as a means of mystifying their art rather than translating it or transmitting it clearly, who use jargon to obfuscate rather than clarify, are subjecting others to Close Encounters with Jargon of the Second Kind, at the very least. They will probably have a difficult time with information systems [1.6.6] that happen to include other people. That covers an awful lot of territory, I know. However, in this sourcebook, and in all your work as a netweaver, you will only have Close Encounters with Jargon of the Third Kind. Won't you?

1.1.2   Technotranslator

There are those who will always work best by themselves, with their specialized tools and reference manuals and specialized languages for company. The work of such persons is important to all of us and may, in fact, result in the production of more jargon. Such jargon may be necessary in order to give mere thought forms ways to be expressed in words. These individuals should not be expected to be able to perfectly translate their specialized knowledge into common language all by themselves. Such a task may be impossible for them. And if *you* are such a person, remember that others may be intimidated by what you appear to know. Try to be aware of the jargon you are using when you communicate with them about your favorite compiler (buzz!) or modem (buzz!) or assembly language (buzz!). Either take appropriate steps to clarify and define what you are talking about or seek the assistance of a technotranslator (buzz!).

> **Technotranslator.** Someone with special talents who can take mere jargon and turn it into pure gold: i.e., useful information for others.

1.1.3   Buzzword(s)

> **Buzzword.** An important-sounding word or phrase connected with a specialized field that is used primarily to impress the layperson.

A buzzword may be jargon and is itself also a buzzword. A recent slap-dash book on micros had the words "buzzwords," "bits," and "bytes" in its title, showing how confusing words can be, since not all jargon (see above) qualifies for buzzword status. A buzzword is Jargon of the First Kind. Whether a particular neologism becomes a buzzword or Jargon of the First Kind depends a great deal on the communications system consisting of the speaker or writer (the transmitter of the message) and the hearer or reader (the receiver of the message). The phrase "What do you mean by that?" may just be the most useful tool for shaping the LANscapes of the Information Age [4.16].

1.1.4   Bugbear(s)

> **Bugbear.** (1) An object of excessive dread. (2) A hobgoblin or bogie.

Our lives are full of bugbears. They are part of the baggage of the human condition. Fear of death. Fear of aging. Fear of technology. Fear of change. Fear of new vocabulary. Fear of loathing Las Vegas. Fear of those who use new vocabularies, or perhaps just excessive awe of such persons. Phobias of subtler scope and less public exposure lurk in every psyche. (Boo!) Fear of keyboards! Fear of mice! Fear of word processing! Most of the time it does not pay to try to demolish other people's bugbears. We each have to exorcise our own goblins. The remedy for most fears is usually quite simple: knowledge, liberally self-administered.

1.1.5 Application Checklist

- When working with people who are just entering the Information Age, or with people who are just beginning to realize that they are already *in* it, use neologisms and jargon carefully and sparingly.
- Learn how to recognize and avoid Jargon of the First Kind.
- When communicating with nonspecialists, carefully define the jargon as you use it.
- Respect other people's bugbears, and try to identify your own.
- If someone else uses jargon and buzzwords you don't understand, don't be afraid to ask what the new words mean.
- Using information to "snow" people or impress them with your expertise may backfire in the long run. And they probably won't like you much, either. You may be forgiven, even if you don't know what you're talking about, if you can make 'em laugh.

1.2 CYBERNETIC ORGANISMS AND ANTHROPOMORPHISM

Anthropomorphism. The attribution of human characteristics, motivations, or behavior to inanimate objects, machines, animals, or any other-than-human natural phenomena.

Calling micros "thinking machines" is a kind of anthropomorphism, since "thinking" is an activity that—presumably—only human beings can engage in. Not to worry. Only the most naive of newsprint journalists still commit this early Information Age faux pas. But the whole thrust of robotics, information science, and micro technology is toward something called "artificial" intelligence [1.8.2] and "expert systems" [1.8.2.4]. In this direction lie machines and systems with ever-more-humanlike characteristics, including the ability to lie, or at least unintentionally misrepresent themselves severely. Which brings up the question: Can we program only those human characteristics that we value and would like to incorporate in machines without evoking other human qualities as well, including ones we may not like very much, such as the ability to tell lies, social and otherwise?

Is anthropomorphism good or bad? Are artificially intelligent machines possible? Better than humans? Equal to humans? Worse than humans? Should we put our trust in machines that make medical diagnoses or predict where oil or coal or uranium might be found or that tell us that the Other Side has launched doomsday?

Cyborg. CYBernetic ORGanism.

Computers, like people, exist in a context that is partly social and partly mechanical. But even the mechanical parts, insofar as they are determined, are determined both by other mechanical factors and by human beings. It seems we are inextricably intertwined with our inventions. To the extent that machines are determined by their mechanical context, they are more predictable in their behavior, more "controllable" than humans. But when you add a human being to the mix of micros and data, you get a system that is both conscious and intelligent, and thereby less predictable, more prone to errors, mistakes, accidents, and disasters.

What about adding a micro to a human being, as in the case of a physically challenged person in an "intelligent" or "smart" wheelchair? Is this a mechanized person or a personalized machine? A human individual who has vital bodily processes controlled by computerized or cybernetic devices is sometimes called a cyborg. Similarly, any system consisting of one or more cybernetic, general-purpose computing machines that requires a human being to control some of *its* vital processes may also be called a cyborg.

1.2.1 The Turing Test

Figure 1.2.1 shows a hypothetical setup for the Turing Test. This test is referred to often in the more esoteric literature of computing, and it accounts for some of the anthropomorphism encountered in the popular press and the tendency of computer neophytes—and even some old hands—to name their micros George or Chip or Schroedinger.

The test assumes that an observer (the Human Being in the diagram) interacts through a terminal with an "entity" of some sort. He can *only* interact through the terminal. (This is just an experiment, not a prediction of how we'll all be necessarily and continually interacting with each other in years to come.) Note that the barrier may be a wall or partition that hides the entity. The entity, for that matter, may be on another planet or in one of Stephen King's "zones." When we use telecommunications networks, the Human Being or entity at the other end may be located in a different part of the country or in another country altogether from the rest of the experiment. The Experimenter (not shown in the diagram)

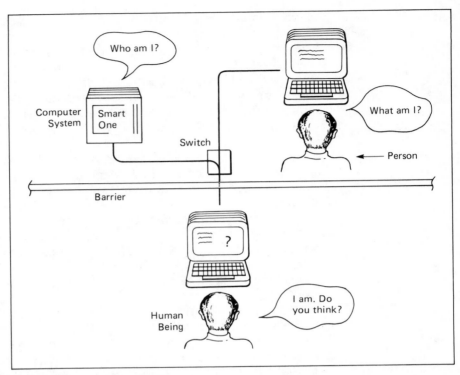

FIGURE 1.2.1. The Turing Test

decides whether the Human Being will interact with the entity—in this case an artificially smart computer—or another person. That's what the switch is for.

Now, Ms. or Mr. Human, put yourself in the place of the subject of the experiment. Using only the terminal, you are allowed to ask questions, make statements, pose philosophical and mathematical problems, and otherwise interact with the entity(ies) on the other side of the barrier. Your mission, should you decide to accept it, is to tell the Experimenter exactly when you are interacting with the entity (computer) and when you are interacting with another Human Being (almost) like yourself.

In essence, the Turing Test claims that if you cannot, using every clever conversational means at your disposal, tell the difference between the machine—if machine it is—and the person, then for all practical purposes the machine may be called "intelligent." It thus passes the Turing Test.

Note that the concept of intelligence itself isn't dealt with in much detail. This is frustrating for humanists, who are fond of attacking the concept of intelligence in machines. Turing certainly never defined the term. The Turing Test is empirical. The machine either passes the test or it doesn't. That, in part, is why there is still such noisy debate about artificiality and intelligence.

One assumption in the Turing Test is that the Experimenter and the Human Beings who participate are themselves intelligent. In fact, they may all be possessed of IQs less than 100 (obviously not netweavers, and only then if you believe that one can "measure" human intelligence), in which case many currently available micros, running the appropriate software, of course, can be regarded (by some) as already more intelligent than most humans *in a given area of expertise or function.* Even for many people of "average" intelligence, the Turing Test has already been "passed" by a number of computers running particular programs. If you confine the interaction to a particular universe of discourse, say chess playing, then machines surpass many Human Beings. This experiential fact has not been lost on the masses and on the mass media, no matter how much linguists and other carbon-based life forms may quibble about just what constitutes "thinking" or "intelligence."

It becomes even more interesting when you consider that, as soon as you enter the electronic networks [1.7.3] you are interacting sometimes with a person, sometimes with "just" a machine, and oftentimes with a combination of the two. After a while, that entity outside your particular barriers blurs into a Vast Active Living Intelligence System (VALIS). At that point, the network of machines-cum-living-individuals (including you) exhibit uncanny behaviors, positive and negative synchronicities [1.10], and other transsystemic qualities that are not only fascinating but useful.

1.2.2   Heuristics

Heuristic methods in problem solving are as old as the hunter-gatherer's rules of thumb for finding food. Heuristic methods *may* lead to a solution of a particular problem. In telecommunications and netweaving, heuristic methods are all we have, currently, as the information environment, hardware, and technology generally are changing so quickly that any strictly formal approach to netweaving would soon be left in the dust of transformation. Heuristics are not 100 percent reliable, effective procedures or algorithms. They are merely plausible ways to begin working on a particular problem set.

The most important heuristic principle, first enunciated in 1945 by George Polya in *How to Solve It* and used by artificial intelligence researchers to the present day is: The end suggests the means. For netweavers, that means using networks to create networks, using information to tailor and re-create new information.

1.2.3   The Cult of the Sacred Cyborg

In the past few years (c. 1980–1984) a sort of religious movement has grown up around the micro. I call it the Cult of the Sacred Cyborg. It goes back many years but has gained momentum recently. This development is not surprising, given the human tendency to anthropomorphize. Some of the enthusiasms of the computopian variety are a welcome antidote to the more popular cult that litanizes, over and over again, its historically redundant observation that Things Aren't Working Very Well As They Are. Nor do I intend to debunk the basic faith that systems—even complex systems—can be changed for the better, however you may define "better." But the tendency to believe that the microprocessor alone is going to solve all our personal, organizational, economic, political, and other social problems should be checked at the door to the Information Age. Having too-high expectations may be worse than having no expectations at all. Disappointment and dashed expectations can lead to early burnout and giving up. The more complex the problem to be solved, the longer it will take to change the systems involved. And we tend to underestimate the complexities involved even in seemingly small systems—especially those we ourselves may create.

The information processing outlook reinforces a certain amount of legitimate anthropomorphism, for it proceeds from the basic view that computers and humans are both instances of the generic "information processing system" or "adaptive system."

The microcomputer by itself cannot do much in the way of changing adult human beings. (On the other hand, there is some evidence that micros plus kids are a different story altogether.) If you have already dealt with choosing, installing, and integrating a micro into your personal livelihood, then you probably won't succumb to the tendency to oversell its benefits to others. If you are even mildly enthusiastic, though, you may be perceived as a True Believer in the Cult of the Sacred Cyborg.

What about networks as messiah? A better bet, theologically speaking, but still oversimplified. The microcomputer—and especially the micro as related to netweaving—changed my life, life-style, and livelihood. I think for the better. You may experience otherwise. Those who point out that the micro is "only" a tool have a valid position: what the individual brings to the tool is as important as what the tool can do. Charcoal in the hand of a Rembrandt will produce results much different from the same charcoal in the hands of a three-year-old.

This last analogy breaks down somewhat, though, when you look at the difference between all previous "tools" of humankind and microelectronic tools. The latter are more universal, applicable to a wider variety of problems and processes. They are more malleable than all previous dedicated tools of the Industrial Age. And, in fact, this tool can transfer some of the intelligence of a Rembrandt (or other great artists and thinkers) into the thought processes of the rest of us. (Cynics might point out that the other trend is also present: the great are reduced to the lowest common social denominator via the mass media and mass transfer of their work. Big MacInfo Burgers for EveryPerson.)

We can be liberated by new technologies, but only to the degree that we are able to accept personal responsibility and discipline. Perhaps only to the degree that we are able to liberate ourselves *sans* machines.

_____ 1.3   TRANSFORMATION AND ECONOMICS

In recent years, the words "transformation" and "network" have been used as generic terms appealing to a wide variety of people taking seminars, personal development programs, workshops, and the like. Thus, for example, Marilyn Ferguson's *Aquarian Conspiracy* uses the word "transformation" to imply many things simultaneously: personal spiritual renewal, "right livelihood," the "collapse of hierarchy," "voluntary simplicity," and so on. Critics point out that "transformation" as currently used by human potential marketeers is so vague as to be nearly devoid of meaning anymore.

Critics of a Marxist bent say that political economics is the arena in which some kind of fundamental transformation is most needed but least likely. They also say, these critics, that John Naisbitt's megatrends (the info-revolt, global interdependency, self-help, self-reliance, and self-transcendence, the stagnation of representative democracy, the emergence of new networks and special interest groups, and multiple life-style options) do not or may not *necessarily* lead to a "culturally enriched, sustainable society." Without going into such nontechnical questions as "What qualifies as being 'culturally enriched'?" and "What's the optimal path to sustainability?," we must take a short tour of political economics.

Since this sourcebook is designed to be primarily practical, our tour must remain brief. But it is necessary so that you and I can use the word "transformation" without sounding like *ex officio* spokespersons for a New Age sales conspiracy, Aquarian or otherwise.

1.3.1   Waves of Change

Borrowing the broad brush of futurist Alvin Toffler *(The Third Wave)*, three historic technological epochs or waves are postulated.

Agriculture was the primary technology of the First Wave. Farming and its attendant surpluses gave birth to the city: urbanization. Begins around 8000 B.C.

Manufacturing (industrial society) is the technology of the Second Wave. The birth of the scientific method. The Renaissance. Begins in the thirteenth century.

Third Wave technology is cybernetics. Since 1945 (Hiroshima; the Atomic Energy Commission; espionage and disinformation; propaganda wars), manufacturing—the Second Wave—has become subsumed in importance by *information products and services.* One typical estimate puts *information*—the "stuff" you are holding in your hands right now— as accounting for a little over 50 percent of the Gross National Product (GNP) of the United States.

GNP figures do not account for the micro-economy represented by households. One writer has called this unaccounted-for sector "Home, Inc." There are many goods and services produced and consumed within the average household, and the emergence of an information economy makes possible even more productivity in the Home Sector. Nor do GNP figures account for the gray-and-black-market economy, where goods and services flow without reports to the Internal Revenue Service or other regulatory monitoring agencies. Information technologies also support the growth of this unregulated sector.

Advances in communications and electronics have been at the heart of the Third Wave. Nine out of ten jobs created in the United States over the past twenty years can be traced to growth in fiber optics, network services, micros, data base construction and maintenance, satellites, etc. In 1970 media/communications services were a $70 billion industry. In 1980 they were approaching $200 billion. By 1990 they are predicted to double again, according to conservative trend analysts, to about $400 billion. We are in the midst of the creation of new wealth on a scale unparalleled—and previously unimaginable—in human history.

1.3.2   Information Economy

Since this new wealth is based on new technologies and new information, there is no reason to think that such wealth *will* or *must* be confined to the industrial "haves" of the planet. Nor is there anything inherent in the underlying technology that must, perforce, create a newer kind of class structure separating the so-called information haves from the information have-nots. The information economy is fundamentally an open system and global in its scope from the outset.

In this book, when I talk about or use the term "transformation," it is to this ongoing, fundamental change that I refer. It is a process that was well under way by the fifties, wrought by our increasing scientific and technological knowledge. It is manifested most obviously in socioeconomic changes. The transformation from an industrially based, relatively closed manufacturing system of wealth generation to an information-based, relatively open system of teledelivered wealth generation is driven both by technology *and* social change, interacting in ways that are not yet fully mapped or understood by anyone. They are least understood by dogmatic partisans of various "isms" and ideologies.

Even though about 60 percent of us now work with information, and even though only about 13 percent of the so-called labor force is now engaged in actual manufacturing jobs, the transformation under way will continue, unless less than 5 percent of our population is needed in manufacturing (primarily cybernetic overseers), and less than 30 percent of the rest of us will be working with information in the traditional ways (teaching, programming, secretarial and clerical work, consulting, and so on). What will the rest of us do? The First and Second Wave answer is: starve. The Third Wave answer is: whatever it was we were doing before we were told we had to "earn a living."

It is entirely possible that ongoing transformations will occur most rapidly at the individual/personal and small-group level, where adaptations to rapid change take place more quickly than at the transnational or corporate levels of organization. Decisions about "cultural enrichment" and "sustainability" may come more and more under the purview of special interest networks and communities of networks. Those who understand the fundamentals of this *information economy,* particularly its *technology* and the successful application of its technology, may accrue new powers and wealth accordingly. Similarly, understanding the overarching concepts that make networks possible will provide links between individual netweavers and planetary macrosystems [1.6.7]. It is also necessary to look at systems [1.6.1] and the nature of systems in order to better understand where and how netweaving fits into this evolving picture.

_____ 1.4   DESIGN

We humans have surrounded ourselves with artifacts of our own design. Design is not some mysterious activity performed only by an elite intelligentsia. It is something each of us engages in practically every day. We make design decisions about how we manifest ourselves and live our lives. Design must not be left solely in the hands of the "experts," whoever they might be.

The design of an information system for oneself, or the design of a network for use by a group of people, or the design of a softspace [1.7] in which to conduct creative practical work, deserves thought, care, and love. Design in nature is lavish and wonderful. It is there to inspire all of us, in full view, right where we may happen to be sitting now. In nature, the most wonderful design for a brain seems to be the one each of us was born with. Whether we can duplicate that design in our softspaces and networks remains to be seen.

Design process. A sequence of creative events that will
(1) improve existing conditions in some way; (2) find clear paths

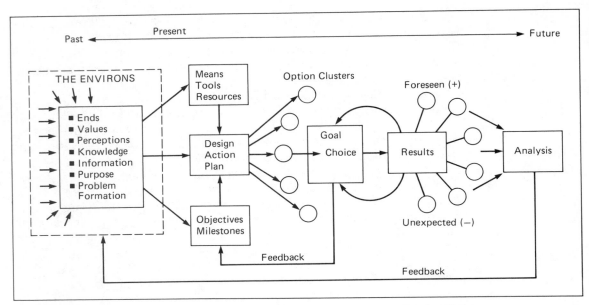

FIGURE 1.4. A Dynamic Systems Model of the Design Process

out of perceived current dilemmas; (3) steer an individual or group or organization toward some future concrete reality; (4) any combination or all of the preceding.

_____ 1.5 MODELS

It doesn't matter where we start, says prominent systems theorist C. West Churchman, in our "search for ways of improving the human condition." All problems unfold into the General Problem. He goes further in claiming that if a given problem does not unfold into wholeness this way, then we can expect "to find that reflection has been 'cut off at the pass.'"

We are not only surrounded by designed material objects of myriad kinds. We are also enveloped in designed processes made up of many social, economic, political, physical, and metaphysical realities. How do we begin to understand them? As time goes on and artifacts multiply, will the result be that the individual is crushed under the weight of an overwhelming accumulation of human culture? Is this the Age of Infoglut, as some social critics have claimed? Is the net result of the Computer Revolution a huge (melodramatic and sinister music here, please) increase in worldwide entropy? Has everything become hopelessly complex? How can one person contribute to or influence the juggernaut of history? To experiment in the real world, in real time, is often costly and dangerous.

In approaching real-life situations, we use models, stored in our heads, which will more or less help us muddle through real situations that confront us, whatever that "real situation" turns out to be. Many toys are scaled-down models of artifacts that will confront the child in us once we have become grownups. The concept of a *model*—an artifact or procedure that reflects, in its form and function, some other and usually larger artifact or procedure—is closely related to the concepts of metaphor, paradigm, map, diagram, simulation, world view, frame of reference, and game.

Model (a more formal definition). A set of symbols with an assigned system of relationships.

The problem with models is that they seem to change more slowly than the situations and real-world entities that they claim to represent. This is especially true of abstruse mathematical models of things such as economic systems, global resources, and networks. And it is n times more true of the models we carry around in our skulls. None of us change our minds often enough.

"Science" is a formal system of relationships as well as a human cultural enterprise. The thrust of science is to build models that will allow us to predict and—perhaps—control the outcomes of real-world events, ostensibly for our own good. Even science as we know it, however, is a social process. As such, most of the pressing problems of our day—ecology, energy, population—are what Thomas Kuhn has called "scientific distractions." These interrelated problems are outside what the scientific community deems as worth its time and enrgy. To tackle these problems requires new links between disciplines as well as new metaphors and models.

1.5.1   Metaphors

A metaphor is primarily a figure of speech. But metaphor goes deeper. A metaphor takes a term from one frame of reference and applies it to another. The creation of a metaphor is an attempt to understand by analogy. Like language itself, which is both a prison and a liberating tool, metaphors can be useful devices and straitjackets on our thoughts. Metaphors are closely intertwined with our *belief systems*. "The microcomputer is a thinking machine." "A network is democratic." "1984 is a State of Mind." In design and solution-creation, an inventory of metaphors is often a useful task to undertake. Such an inventory will suggest areas in which metaphors are lacking and will outline the everyday contours of the belief systems and paradigms being brought to bear on the project at hand.

1.5.2   Memes as Metaphors/Metaphors as Memes

Evolution proceeded from molecule to cell. The appearance of the gene—the basic unit of biological evolution—was a milestone in the packaging and transmission of information. It all happened first in the hypothetical organic soups of prehistory. If machines become intelligent, they will do so within a rich new soup: the soup of human culture. The new replicator, the cultural equivalent of the gene, conveying the idea of a *unit of imitation,* is the meme.

Memes propagate themselves by jumping from brain to brain, group to group, machine to machine, like fleas—although they do not bite everybody. Some persons seem immune to new memes, others seem too prone to every mutant meme that happens along. What or where is a single unit meme? Mass broadcasting media in all their forms foster the propagation of memes. Memes must compete for time in the human brains in which they live. They also compete for advertising space, media time, residence in computerized data bases, and so on.

Richard Dawkins, who coined the word, conjectures that memes and co-adapted meme complexes evolve in the same way as co-adapted gene complexes. Selection will favor memes that can exploit their cultural environment to their own advantage. The cultural environment, our collection of artifacts, ideas, metaphors, and paradigms, consists of memes under constant selection. Who or what selects? Each of us. You. Me. And Aunt Hattie. This selection process is an ineluctible responsibility and will contribute to the health or sickness of our bodies, our organizations, and our planet. Memes can be deadly viruses. Our beliefs can kill us. The future begins as a meme in a single imagination. The idea of a network is a meme. The idea of a meme is itself a meme—a metaphor drawn from evolutionary theory and applied to the realm of human creative effort. Any physical artifact is a meme in concrete form.

1.5.3   Icons

> **Icon.** An image, representation, or symbol that stands for something else, usually something physical in the real world.

Icons can be drawn from everyday objects, such as the microcomputer, which itself has become an icon of sorts. Since microcomputers are malleable universal tools, each group that comes in contact with and begins to use micros will tend to put its own memes in the machinery. The micro-as-icon will symbolize publishing to publishers, medicine to doctors, an audiovisual aid to teachers, a graphics tool to architects, a fast tax-preparer for accountants. Icons take up residence in computers and propagate themselves throughout various networks. Icons are a class of meme.

An icon is usually a visual metaphor. A file cabinet on a video screen can represent a micro's memory. A wastebasket on the screen can represent the process of deleting or "throwing away" information from the data base. Icons, being a particular kind of meme, must be constantly evaluated for their life-enhancing or life-reducing potential. Poorly chosen icons, or icons that perpetuate obsolete methods, can choke further development of thought, intention, and organizational evolution.

An iconography of one's personal thought-space is not only useful but imperative in the design of optimal information systems, softspaces, and networks. For example, with newer, higher-powered micros such as Apple's Macintosh, it is possible to create your own visual metaphors for work you do. So if you think of your entire physical space as a source of icons, you might create a file labeled with an icon or image representing, say, a particular storage location (a trunk, bank vault, safe deposit box, bookshelf, and so on). Or you might create icons representing classes of information: the image of a telephone to represent telephone information, the image of a microcomputer to represent microcomputer information.

A planetary iconography is evolving slowly. Example: universal graphic symbols for public transportation systems in various countries speaking different languages. The icon for the telephone is known worldwide. An iconography of networking and microcomputers is also evolving. Figure 1.5.3. shows a few examples.

1.5.4 Games

A game can model aspects of the real world. Parker Brothers' Monopoly is such a game. Among children, a culture of games has perpetuated itself over thousands of years. A game can be a kind of metaphor for teaching new concepts. Or it can reinforce deadly memes. Games can be used to teach new skills, entertain, or distract, sometimes all three at once. New games are invented constantly. Games based on the micro's capacity for simulation and modeling are proliferating. The design of games—or anything else—can be approached as a game.

Game theory is the formal, logical study of decision making. It is especially useful in situations where the intentions of other people are uncertain. Game theory was developed first in the early 1920s by John von Neumann. He and Oskar Morgenstern collaborated on *The Theory of Games and Economic Behavior* while the two of them were at Princeton's Institute for Advanced Studies. Their book was published in 1944 and was intended for economists and social scientists. However, the military quickly picked up on the theory because of its emphasis on and

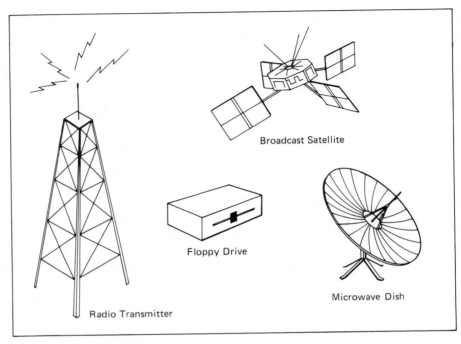

FIGURE 1.5.3. Some Icons Used in Netweaving

rigorous treatment of the concept of *strategy*. One of the evolutionary offshoots of this particular meme complex was the 1983 movie *War Games*. (More peace games, please, game designers!)

The usefulness of games as teaching/learning tools has just begun to be explored by microcomputer programmers and other artists. A big advantage of a game, as with any model of a real-world process, is that it can reduce complexity in order to increase overall understanding.

1.6   LIVING SYSTEMS

R. Buckminster ("Bucky") Fuller presides modestly and majestically over the Information Transformation Revolution Evolution Networking Consciousness Systems Movement and Antijargon Society, now and for aeons to come. Actually, Bucky was never "anti" anything so much as he was an optimistic synthesizer of Magnificent Memes and Meme Clusters. Bucky the Global Shaman told us that a system is "a configuration that divides the Universe into (1) everything outside the system; (2) the system itself; (3) everything inside the system." Which system? The system at hand. You get to define "what" the system is. Just remember that it consists of things inside, things outside, and the system itself. The human being as system: you, your environment, and your insides (including your nervous system, which is one of your major *sub*systems).

> **Living systems.** Complex structures that carry out living processes . . . identified at seven hierarchical levels—cell, organ, organism, group, organization, society, and supranational system. Systems at each of these levels are in turn composed of subsystems that process inputs, throughputs, and outputs of various forms of matter, energy, and information.

The general direction of evolution is toward greater complexity. Evolution is thus anti-entropic. Anti-entropy is also called *negentropy*.

There are, of course, other than living systems. When it comes to defining just what a system is, there are many ways to do so. The minimal definition of system seems to demand the following elements:

1. A set of *states*. Each of these states must be individually identifiable and unambiguously distinguishable relative to the other.
2. One or more *transformations* defined for at least some of the states or sets of states.

A "softer" definition of a system stresses:

1. Many constituent elements having some property in common.
2. A structure—that is, recognizable relationships among the elements. These relationships cannot be reduced to mere accidental aggregations of elements.

Microcomputers lend themselves to systems modeling of all kinds, including the modeling of communications networks.

1.6.1   Systems Theory

In theory and in principle the transformation from an agri-industrial to an information economy can have a positive impact on global living systems in the following concrete ways:

1. Every inhabitant of the world can be adequately fed, clothed, and sheltered.
2. Sufficient medical care can be provided for everyone on the planet.
3. Education and self-actualization can be experienced throughout the lifetimes of each individual on earth.
4. War can be outlawed and appropriate social sanctions brought to bear to prevent the outbreak of illegal war.

5. All planetary society can enjoy freedom of opinion, speech, belief, and freedom of action and travel, thus minimizing the constraint of the individual by societies, economics, or governments.

6. Energy can be essentially free for the asking.

So what's holding us back?

When you postpone thinking about something too long, then it may not be possible to think about it adequately at all.

—C. West Churchman

The most spectacular achievements of the systems approach have occurred—are occurring—in the exploration of space. The 'droids and cyborgs we send to the outer planets are parts of our living systems for as long as they remain in radio communication with us earthlings.

Systems are composed of subsystems that work together for the overall objective of a whole or *the* whole. Thus, systems approaches to various problems and goals are often called "wholistic" or "holistic." Holistic health. Holistic information. Holistic therapy. Holistic medicine. Holistic computing. Sometimes, in their most watered-down forms, these approaches become mere lip service, silly and thus rendered practically useless. The systems approach may disturb typical mental processes, especially those of specialists and licensed professionals. The systems approach requires that one jettison old memes and metaphors that are not working in favor of systems metaphors that may just do the job(s) in mind.

Sometimes the word "system" is used in a buzzwordy way, as in "Down with the system." Which system? Or as in "computer system." Just the computer? Or the computer with accessories and data? How about software? Those who use the systems road to design and problem solving usually define *which* system they're talking about. Ask: What are the components of the system? What are the relationships between the components? What are the inputs of the system? What are its outputs? What are the relevant environmental factors affecting the performance of the system? These kinds of questions can be applied to any system, at any hierarchical level.

All living systems can be analyzed from the point of view of each of nineteen "critical subsystems" that have been identified by James Miller in *Living Systems*. We won't list all nineteen of them here because we will not be interested in all of them in this sourcebook. Rather, our emphasis is on those subsystems specifically concerned with information and information processes. For a complete overview of Miller's nineteen critical subsystems as identified by Miller, see *Living Systems*.

Reproducer: A subsystem that processes both matter-energy and information which is capable of giving rise to other systems similar to the one it is in. Example: a part of a network devoted to creating other networks.

Boundary (membrane, interface): A subsystem that processes both matter-energy and information. A membrane is at the perimeter of a system and holds together the components that make up the system, protects them from environmental stresses, and acts as a "filter" to exclude or permit entry to various sorts of matter-energy and information. Example: The surface of your body is the boundary between you and not-you. Example: Your dwelling is a membrane letting in air, light, and physical objects, keeping in heat, and putting out waste. Example: The communications subsystems of a company form a membrane letting in some kinds of information and keeping out other kinds.

Distributor: A subsystem carrying inputs from outside the system or outputs from its subsystems around the system to each component. Example: the mail department of a company. Example: a company's PABX system. Example: a note from Mom to the whole family pasted on the refrigerator with one of those wonderful little yellow pasties. (In this case, Mom was the originator and the distributor.)

Converter (conversion; translation; transformation): A subsystem that changes some of the inputs to the system into forms more useful for the special processes of that system. Example: the conversion of financial data from numbers into graphs and charts. Example: the conversion of electrical impulses into voices or text.

Input transducer: Sensory subsystems that bring information into the system, changing the form of its markers if necessary so that the system can handle it internally. Example: your eyes and ears. Example: a newspaper reporter.

Internal transducer: A sensory subsystem that receives, from subsystems or other system components, information about significant alterations in those subsystems or components, changing them, if necessary, to other kinds of information that can be used within the system. Example: the Government Accounting Office (GAO) monitors various government subsystems, converting its data into reports that can be fed back into those same subsystems, hopefully improving their performance. Example: the thermostat in your home heating system monitors internal temperature and changes this information into signals that control the heat source. Example: In your body, you have sensitive monitoring systems for your blood sugar levels, internal temperature and hormone levels.

Channel and net: A subsystem that transmits information to other parts of the system.

Encoder: A subsystem that translates interior private codes (jargon) into public codes that can be interpreted by the system's environment. Example: A technical writer takes a company's internal technical documents and converts them into manuals and instructions for end-users. Example: A modem [4.8] takes the codes of one system—your microcomputer—and changes them into the codes usable by another—a remote mainframe computer.

Decoder: A subsystem that alters, if necessary, the code or protocols of the information given to it through the input channels or internal feedback into one that can be used internally by the system or by individual subsystems. Example: Again, a modem does this kind of conversion on the input side.

Associator: A subsystem that carries out the first stage of the learning process, forming enduring associations among items of information in the system. Usually, this associator function is carried out by humans. However, in the design of "expert systems," the associator is partially or fully automated and built into a machine. Example: a computer-based medical diagnostic program. Example: the records management or information management department of a company.

Memory: A subsystem that carries out the second stage of the learning process, storing various sorts of information in the system for different periods of time. Examples: floppy disks, your biological memory, videodisks.

Decider: An executive subsystem taking information inputs from all other subsystems and transmitting back to the subsystems information that coordinates the entire system. Again, in most cases the decider or decision maker will be a human being. Many decisions, however, are of a "routine" nature and can, in theory at least, be programmed into a machine. Example: the policy-making board of a company or organization.

Output transducer: A subsystem that puts out information from the system to the environment. Example: a public relations, publications, or communications department within a company.

While Miller has rigorously defined and separated these critical subsystem functions, in practice two or more of them will be carried out by multipurpose machines or persons.

1.6.2   Channel and Net

In telecommunications, a particular physical path between two devices is called a channel. If there are many channels to many devices or points, then the system of channels is called a network or just a net. There are many ways to configure channels and nets [4.17.3]. With some of the new information technologies, a single channel can link multiple devices easily and economically. This is essentially the advantage of technologies like digital packet switching [4.4.2] and fiber optic transmission [4.23.5]. A particular channel has associated with it a bandwidth [4.5.2], which can be thought of as the size of the "pipeline" that carries information. A given channel or set of channels can also be thought of as a *conduit* [1.7.5].

1.6.3   Holons and Hierarchies

Each of us is a marvelously designed living system. We contain within ourselves subsystems of equally marvelous complexity. Each of us is a part of a larger set of wholes, or sets of wholes. We are fields within fields within fields.

In *The Ghost in the Machine,* and later on in *Janus,* Arthur Koestler used the term "holon" to specify any autonomous system that contained within itself subsystems and that was itself part of a larger whole. A holon transcends looking at the relationships between mere parts and wholes in favor of a both/and model: a given holon is a "part" when looking "up" the hierarchy to the upper-level systems, and is a "whole" when looking "down" the hierarchy to the holon's constituents.

Since the term "hierarchy" has justifiably unpleasant overtones of bureaucracy, and since this particular model of Koestler's can have reorganizable and semipermeable divisions with countless feedback loops—laterally, upward and downward—according to flexible strategies rather than fixed rules, Koestler gave us the word "holarchy." Holarchies are composed of holons. The holarchy proposed by Miller is a living systems sort of division: cell, organ, organism, organization, society, transnational global system. But why stop with the cell? Below the cell are other, more subtle holons: organelles, molecules, atoms, subatomic particles. And further, above the holon that is Mother Earth is the universal holarchy of planet, solar system, constellation, galaxy.

1.6.3.1 SELF

Consciousness is a form of information processing. The individual living human being is the essential system to begin with in any netweaving effort. In systems composed of many people and processes—business, government, education, manufacturing—the individual must be the focus of all analysis, process, and feedback. For the netweaver, this means a

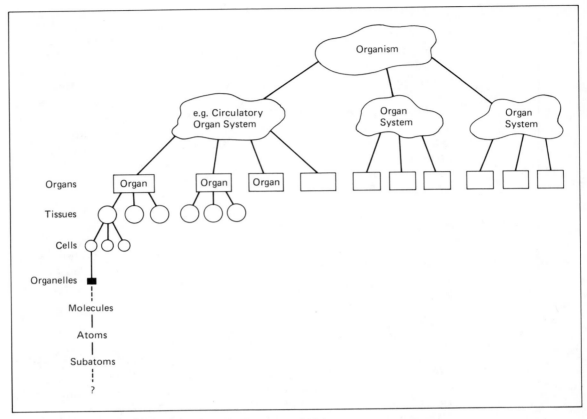

FIGURE 1.6.3. An Example of Holons

certain kind of self-centeredness. A willingness to continually monitor one's own processes and accept the feedback of others is essential for anyone who would presume to co-design networks and network-based organizations with others.

Figure 1.6.3.1 (Human Sensory Channel Capacities) shows the human being's primary information inputs. Notice that we can hear faster than we can speak, a fact that can be used to good advantage in personal information processing.

There are, of course, essential differences between biocomputers and silicon-based microcomputers. There are also similarities, concepts and models that apply to both. Computers can store masses of data accurately and retrieve same with little difficulty. The brain's memory is not as good. In fact, a kind of basic inexactness seems to be built into the design of our brain. But this "inexact flexibility" is precisely the brain's

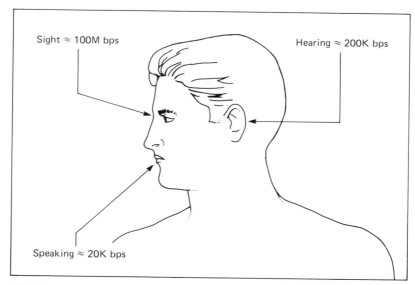

FIGURE 1.6.3.1. Human Sensory Channel Capacities

strong suit. We are unsurpassed at making guesses, making shrewd connections, and discerning/creating patterns where none were before.

The design of the brain is that of a communications device rather than a steam engine. It is a system for coding and organizing information. The brain is not a passive receptacle of messages sent to it from the environment. It maintains itself as an information system in more or less "stable" balance. It selects and structures. It switches on and switches off. At any given time, as much as 80 percent of the neurons in the brain may be acting to suppress activity rather than produce it. Selective attention is the hallmark of the human nervous system. Each act of perception at each moment is unique to that moment.

It is useful to look at biocomputers as information-processing systems supported by the rest of the body, but operating somewhat independently of it. The senses provide input. The biocomputer then responds in many different ways. One kind of response is evident in so-called reflexive behavior. The effect is that information is transferred with minimum delay from the afferent channels to the nerve fibers controlling effector muscles. The biocomputer is also a clearinghouse for internally sensed information. This information plays a role in controlling physiological processes. Yet another biocomputer function is to render information into patterns that orient us in our sensory environment. Our pursuit of security and/or gratification is initiated and controlled by primal biocomputer programs.

1.6.3.2
GROUP

Groups are a dominant feature of personal and organizational life. A group, usually a small group of from three to twelve people, is the holon linking the personal and the organizational, the personal and the planetary. Groups dominate organizational and institutional life.

Groups can be called by many names: team, task force, committee, commission, board of directors, network. A basic knowledge of group dynamics can help a netweaver accomplish tasks within an organization or special-interest group. Although there have been many criticisms of group behavior in years past, especially for the mediocre quality of groups' products and their adverse effects on individualism, the fact remains that many things that humans view as desirable cannot be achieved by a single individual working alone in a small room with nothing but books and finely tuned sensibilities. Groups function pretty much as their members make them function. Work accomplished in the context of a group can be just as exhilarating, satisfying, and remarkable as work accomplished by individuals. Poor design can also result. The issue, in any given case, is twofold: Should this be done by a group? If so, how? The meta-issue of the *design of the group and its interactions* is a major concern of the netweaver.

Most of the important decisions made in systems of all sizes are made in settings that possess all the dynamics of group problem-solving sessions. A casual conversation between colleagues that leads to a decision, a chance exchange of information between boss and subordinate, a spontaneous coffee-break discussion of pertinent issues: these are all group decision sessions. This kind of "serendipitous" group activity is greatly increased using the new telecommunications media. The network setting brings people together even though they may no longer be seeing one another at the water cooler or coffee-break station.

1.6.3.3
ORGANIZATION

What may be most in need of innovation is the corporation itself. Perhaps what every corporation (and every other organization) needs is a department of continuous renewal that could view the whole organization as a system in need of continuous innovation.

—John W. Gardner, *Self Renewal*

New organizations are invented in the United States on a daily basis. Tocqueville noted, in his classic study of America and Americans, that our democracy tends to give rise to all manner of peculiar and useful affinity groupings and organizations designed, ostensibly, to give a voice to the voiceless, power to the powerless. Old organizations are called institutions. Old institutions are often criticized for their rigidity, resistance to change, stultifying and alienating working conditions, and unconscious and unconscionable oppression of the innocent. Every year, pollsters query the Great American Public to find out how much faith the people still have in various institutions. Congress is only one of many institutions that wax and wane in these popularity polls. Government itself is institutionalized to a degree unforeseen by the founders of our country.

At its worst, organizational life is characterized by the negative term "bureaucratic." Yet, at its best, organizational life can accomplish great things. Again, these accomplishments are usually not of the kind that can be achieved either by an individual working alone or by a single small group. The nature of the tasks undertaken by large organizations is perforce complex. The problems of the organization—whether related to production, design, or management—are usually systems problems.

The bureaucracy as an organizational form was perfected during the Second Wave, the Industrial Revolution, to organize and direct the activities of industry, government, education, investigation, religion, and voluntary associations. Developed under nineteenth-century conditions, bureaucracy was probably an appropriate sociotechnical response for its time. However, under conditions of constant change and transformation—essentially the conditions for some time to come in the Information Age—the bureaucracy as a social invention now appears inadequate to many of the challenges of human survival.

Netweavers should look for the following telltale signs of bureaucracy in the organizations they come in contact with:

1. A well-defined chain of command with few, if any, horizontal links.
2. A system of procedures and rules (usually set forth in an "operations manual" complete with chain-of-command organization chart) that attempts to deal with all contingencies relating to work activities.
3. A high degree of labor specialization, with each isolated worker having little, if any, idea how the organization as a whole works or what role(s) others play in it.
4. Promotion and selection of individuals based solely on technical competence.
5. Impersonality in human relations.

A bureaucracy is typified by the hierarchical pyramid arrangement seen on the organization charts of many different kinds of institutions. The lines connecting the boxes on an organization chart are lines of communication and authority. The model on which it is based is essentially a machine or mechanical model, not an electronic or cybernetic one. It was developed as a reaction against the nepotism, personal subjugation, cruelty, and subjective arbitrary judgments that passed for "management" in the early days of the Second Wave. Second Wave organizations needed order and precision and predictability. The workers needed fair and impartial treatment. This military and ecclesiastical form was ideally suited to the needs and value systems of the Victorian era. Today, however, new designs must be developed.

Netweaving is a key part of a new approach to organizing for human purposes. Organizations are holons that contain groups and individuals. Organizations are components of a global economic system. Organizations-as-holons must first be transformed into networks and network-based structures, with organic cybernetics as the main organizational model. Long before the industrialized countries can complete this transformation, Third World countries can "leapfrog" into the necessary structures and join the world Information/Energy grid fully prepared to hold their own as information *creators* and thus as information "haves."

**1.6.3.4
ORGANIZATION
DEVELOPMENT**

Organization development (OD) is one possible response to change and the need for change. It is best described as an educational strategy—formidable though it may be—aimed at changing the beliefs, attitudes, values, and, if need be, the *structures* of organizations, both old and new, so that they can better serve not only their own participants, but society and the planet as a whole. Here, better service means a survival orientation, as in survival of the species called *Homo sapiens.*

OD as a field of study is about twenty-five years old. It is concerned with such things as team development, intergroup conflict and conflict resolution, confrontation, feedback, and planned organizational change. A basic value underlying all such OD work is that of *choice.* Better decisions are likely when there are real choices to be made. Freedom, whether personal or organizational, is correlated with the number and quality of choices to be made.

**1.6.3.5
GLOBAL ACCESS
TO THE
WORLD GAME**

Global Village. Spaceship Earth. Whole Earth. Mother Gaia. Certainly the *awareness* of our planet as a whole living system is in the air, the stuff of Sunday supplements ever since the first space-made photograph of our Beautiful Blue Pearl. But, as with any therapeutic insight, the gap between awareness and behavior is still huge. The question remains, in Jack and Jill Everyone's mind: What does it all mean? Can this insight alone dissolve the old bureaucratic approach to global problems and opportunities such as war, famine, disease? Where will "global consciousness" reside and how will it act? The World Game model suggests a new approach.

The concept of World Game came from R. Buckminster Fuller. Rather than play "war games," Bucky thought, which is essentially a zero-sum game, why not play a game in which no one loses? A non-zero-sum game for all humanity? Why not turn our creative activities from weaponry to livingry? Well, why not? The World Game is simple in concept, complex in implementation, as one would expect. Its techniques and premises, however, can be applied on local personal and organizational levels, as well as on a global scale.

The fundamental premise of the World Game is that one can begin to anticipate the future by making a thoroughly detailed inventory of the present.

Here are the steps to a World Game. First, inventory the entire earth's resources, physical and metaphysical (ideas, information, meme pool). Then inventory the needs of everyone on the planet. Finally, design strategies to meet those needs using the available resources—abundantly.

Simple, yes?

1.6.4   Cybernetics

Cybernetics is the science that will finally undo the bureaucrats. It is the science of communication and control. It is useful in studying living systems behavior as well as the behavior and control of machinery. It is profoundly philosophical and at the same time practical. It is useful in the work of developing organizations that can meet the changing conditions for continued survival, and in designing totally new organizations to meet unique, evolving needs.

Any system that *responds to its environment* is a cybernetic system and can be studied profitably from that point of view. And by "profitably," I do not mean simply metaphysically profitable. Any action leading towards a goal must be controlled in order to achieve that goal. Cybernetics teaches how. Progress toward a goal can be known only through some form of communication. The process of constructing networks for oneself or others is essentially a cybernetic process. It is also concerned with communications systems in larger holons.

Figure 1.6.4 shows a simple model of a cybernetic process with which most Americans are familiar: the thermostat on a heating system. Part A, with its accompanying graph, shows the process behavior in an "open loop" system. (Really, an "open" loop is no loop at all!) Part B shows a closed loop configuration, which is the typical one. In some milder climates, where central heating is used infrequently or sparingly, if at all, the open loop system of part A can be made into a closed loop system simply by putting a human being—equipped, after all, with exquisite temperature sensors and controls of its own—into the system to "close the loop." When it gets too hot, the human opens the switch and turns off the heat. When it gets too cold, the switch is closed. The graph of the behavior of this elementary "cyborg" then approximates the graph shown for part B.

In a large social system that includes one or more networks, various subsystems—perhaps departments or individuals—will monitor the

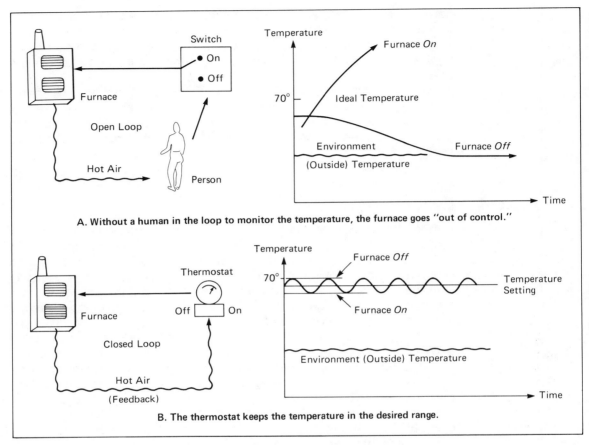

A. Without a human in the loop to monitor the temperature, the furnace goes "out of control."

B. The thermostat keeps the temperature in the desired range.

FIGURE 1.6.4. An Example of a Cybernetic Process

functions of other subsystems and of the organization as a whole. Feedback from this department then becomes the "control signal" for the target subsystem. Air pollution, for example, is monitored by a "pollution thermostat" in the form of an air-quality control board. When the pollution level rises beyond a certain point, the "thermostat" puts out warning signals on the various communications networks. If a particularly bad source of pollution is pinpointed, the "thermostat" can zero in with a signal to the source to desist its repulsive output.

1.6.5   Information

To live effectively is to live with adequate information.

—Norbert Weiner, first cybernetician

The whole system called universe is based on chance, not accident, so we are told. Chance *(yang)* and antichance *(yin)* coexist in a complementary relationship, joined in necessity. The agent of chaos is called entropy: the random element. The nonrandom element is called information.

Information takes the uncertainty of the entropy principle and creates new structures, in*form*ing the world in novel ways.

It wasn't until the 1940s that "information" became a scientific term. No other single concept links so many diverse ideas with such great power: ideas in linguistics and biology, psychology and philosophy, computer science and sociology, art and probability theory.

Today, we think of information as "news," "intelligence," "facts," "knowledge," even an "industry." The mass media bill themselves as "news *and* information networks." In medieval times, the word "information" was used more actively to suggest something that gives form—in-forms—to material substance, or character to the mind. Information was a force that shaped behavior, trained, guided, instructed, or inspired human beings. After it was defined precisely enough to be useful to telecommunications engineers, the word began to regain some of its medieval connotations. Information *is* an active agent. DNA's message gives shape to the cell, organ, and organism. The message from a radio transmitter guides the orbital path of a robot vehicle on its way to the wastes of Mars.

Information is a universal principle, active in the world. It spans computers and physics. It is important in molecular biology and human communication. It continues to be useful in understanding the evolution of languages and the evolution of living things.

Nature = matter + energy + information

Our species has been studying matter since the dawn of history. Energy and its interchangeability with matter began to be understood with the arrival of Einstein. Information arrived on the scene in 1940, and we have only just begun to feel the effects of the theory of information.

Information is the "stuff" of the information economy, the Information Age, the micro revolution. Information shapes networks, organizations, and human character. Sometimes, information shapes itself. To study information is to study symbols, signals and noise in biological, mechanical and biomechanical systems.

Information is a *strategic resource.*

**1.6.5.1
SYMBOLS**

One definition of the human being is "the animal that makes symbols." We are immersed in a sea of symbols that convey information. Generally, symbols are distinguished from "signs" or "icons" in that symbols bear an arbitrary relationship between themselves and the meaning they convey. Symbols are created by agreement. The alphabet is a set of symbols used in many languages. A sign or icon, on the other hand, usually bears a visual relationship or "mapping" to the object(s) of the sign. A picture of a person walking and another picture of a hand shown palm out, used as a set to control pedestrian traffic at an intersection, are signs or icons. They are understandable across linguistic barriers. The words WALK and DON'T WALK, on the other hand, are sets of symbols. They are understandable only if you happen to speak English. The memes associated with both the symbols and the icons are identical. However, the memes are expressed in ways that are very context-dependent. These distinctions are increasingly important to netweavers. For that matter, they are important to anyone who will use a microcomputer over the next ten years. Universal sign languages are beginning to evolve in the micro world, where icons understandable across linguistic barriers are being developed. We see the first hints of this trend in imported automobiles of the 1970s. LED outlines of gas pumps, thermometers and seat belts communicate the machine's condition to the driver. Increasingly, as world markets for microcomputers and micro systems expand, the use of icons rather than symbols will take root. A planetary hieroglyphic is evolving. Rather than being the development of theoreticians who long for the day of Esperanto, this universal hieroglyphic is growing in many places simultaneously. Conceivably, netweavers and user communities and other communities of consciousness will be primary contributors to the evolving hieroglyphic.

1.6.5.2
SIGNALS

A signal is any varying physical characteristic deliberately shaped (modulated) by a transmitter to carry information to a receiver. In telecommunications systems, signals are usually electronic. Signals can be transmitted by light waves, radio waves, or sounds. By varying frequencies. By changing from presence to absence and back again. By subtle changes in duration. Signals can be transformed from one way of impressing information to another.

1.6.5.3
NOISE

We all have experience with noise in many different kinds of signals. Our biocomputers manage to filter a lot of it out. Our perceptions have a lot to do with just what is defined as noise in any given instance. The noise of the seashore is detrimental to the interior of a recording studio located on a beachfront. The lull of the waves is soothing to the owner of a beach house at Malibu.

Noise appears on our television sets and we call it "snow," because that's what noise looks like in that medium. On our radios we call it "static," as in "Don't give me any." On our telephones, noise shows up as crackling, fading, whistling, or crosstalk.

Noise. Any undesired alteration of a signal with the potential for obscuring meaning.

1.6.6   Information Systems

An information system is a cybernetic system that assists humans in responding to their environment. An information system responds to and makes changes in the information or media environment. An information system can also respond to and make changes in the physical environment. Increasingly, information systems are changing the physical ecology of our planet as well as the information ecology—the sea of memes in which we dwell—of our culture.

Prior to the Information Age, information was power. Information is not power. In the Information Age, nearly everyone will have nearly all the information all the time. Significant differences in access will even out over time, because information per se is essentially uncontrollable. The measure of power now is the extent to which information can be applied, how much it can be *used*. Not how much it can be *sold* for, nor how many times it can be sold, but how much it can be *used*. A corollary of this is that power will be accrued according to how close one can get to the sources of *usable* information. Better, still, to *be* the source of *new* usable information. What is usable, you will note, is up to you to define. Only you can decide what you can use. And you will be sold information you *can't* use until you can learn how to tell the difference.

An information system is a living system looked at from the point of view of information technology on the human scale. An information system is truly living, by virtue of human participation as designers, co-designers, and/or as beneficiaries (or victims, as the case may be) of the system. In many information systems, a human being, or groups of human beings, are part of the cybernetic loop, altering the output of the system according to human criteria and considerations that only humans can have. An example of this sort of cyborgian integration of persons and machines is the military NORAD system, in which, regardless of the information presented by the computers, only the president of the United States can actually "push the nuclear button." At least in theory.

The essential task of the netweaver is concerned with information systems and the interlinking of information systems into larger holons, up to and including planetwide networks. An information system is composed of people, data, hardware, and software. An information system may or may not be part of a network, but all networks are information systems. The network may only be a temporary, two-node system. The essential ingredient that turns an information system into a network is its *connections*.

In netweaving, the essential connections are usually those between pieces of hardware in a telecommunications network. However, other kinds of connections may be equally important to consider when setting up a particular information system or network. Friendships, colleague relationships, and overlapping interests or concerns also form the basis of connections between people. In addition, there may be connections between different kinds of data and data structures in an information system. Sometimes it is important to connect software as well as hardware. In an information system of even modest size—say, your micro and its local peripheral equipment, you, your interests, and your friends or workmates—the combinations of relationship and connection can become complex indeed. Careful attention to each of the components of an information system can usually unravel these connections and render the system more useful and usable thereby.

1.6.7   Noöspherics

Noöspherics is the study of global consciousness. The fact that no one knows what global consciousness is, is no deterrent to such study; it is, rather, the essential point of such study. No one really knows what *individual* consciousness is, either. The existence of consciousness is both doubted and denied in some circles, even in this enlightened age. Yet there is a growing community of people studying consciousness, and an even-faster-growing body of information available on the topic.

Teilhard de Chardin coined the term noösphere to denote the organized layer of intelligence on the planet. The lithosphere is the layer of matter on our globe. The biosphere or ecosphere is the interlocking system of life forms on the planet. Information systems and networks compose part of infrastructure of the noösphere. The information flowing through and between networks helps orchestrate the thoughts of the global brain. Given how youthful this development is, historically, we should not be surprised that these "thoughts" are, on the whole, chaotic, contradictory, and conflicting. Mother Gaia is just entering her adolescence.

1.7   SOFTSPACES

Softspaces contain and are contained by information systems. The two intertwine, as it were. Softspaces are "soft" because they are characterized by concentrations of machines, energy, and information. The information "contained" by a softspace may reside physically within the walls of a given micro installation. While the emphasis is on its physical aspects, a softspace can reach out to include lots of information that may reside elsewhere, in data banks and biocomputers. A softspace is oriented toward the effective use and manipulation of information by people, whatever its content or presentation form (visual displays, hardcopy, audio, etc.). Your softspace is that place where your information system hardware is installed. Its "center" is your body/brain/mind. Whereas an architectural space is defined by physical artifacts and measurements, softspaces are composites of personal, architectural, and information spaces. Design of a softspace includes considerations such as lighting, heating and cooling, and the comfort of the body. These are *ergonomic* considerations [2.5.3].

Design of a softspace also includes those things that designers of information systems and networks are concerned with: user friendliness, ease of access, appropriateness or "fit" between needs/goals and information. A softspace may include a musical component (stereo and hi-fi), a print component (libraries of books) and a video component, in addition to the micro and its software. The term "mediaspace" is sometimes used to describe the installation of a softspace in a business or institutional environment. A good softspace is both comprehensive and comprehensible. The "office of the future" is a highly portable personal softspace capable of interacting with a wide variety of other softspaces (networks, data banks, media, people).

An essential component of a softspace is *two-way communication.* Mediaspaces, on the other hand, are often concerned only with the installation of one-way (mass, broadcast, noninteractive) media, whether text, visuals, or audio.

In the hard jargon of the micro engineers, "end user interface" designs are also part of the way a softspace comes together. The design of keyboards, screen displays, touch-sensitive surfaces, "mice," and the interactive levels (or "presentation level") of software all determine the function and "feel" of the boundary between people and information processing machines. With the installation of enough microprocessors in enough places, softspaces will merge with the ecology of local and planetary environments, creating cybernetic forests made of electronic and organic trees.

In the Industrial Age, we lived in physical spaces connected by modes of transportation. In the Information Age, we live increasingly in softspaces connected by telecommunications, where distinctions between "artificial" and "real" become more and more arbitrary.

Essential features of softspaces (not a comprehensive list):

- A display of some kind (small, medium, large) that can link to the digital networks using ordinary telephones and lines.
- User ability to enter data, query the remote data base, allow program development, create messages, send and store messages.
- At least one megabyte of data storage capability.
- Ability to connect to many kinds of networks locally (LANs) and remotely. It will generate a variety of network protocols and speeds, depending on need.
- Availability of a variety of peripheral devices.
- Capability of many different kinds of local processing.

In the not-so-far-distant future, personal and organizational softspaces will include the ability to do remote computer-aided design (CAD) of one's own electronic chips according to personalized needs and specifications until enough information is acquired to enable the chip to be made. The resulting design will be automatically sent to computer-aided manufacturing (CAM) centers for production of as many units as are called for. The design for the chip will be placed in a central archive in case others want a chip with the same features.

The initial design of your softspace, and all potential electronic additions or expansions thereafter, should include as many of the above functions as your budget will allow. As time goes on, the cost of meeting these general yet powerful specifications is coming down, and we will certainly be adding to the list of functions. In the meantime, there are ways to maximize your info-investment dollar per unit of info power and still begin meeting many of these criteria.

When can you have one, you ask? As soon as you're ready, or when you demand it, whichever comes first.

1.7.1 Membranes

The boundaries of a softspace are best described as membranes. Just as the wall of a cell admits certain molecules and prevents others from entering, a softspace is also semipermeable. Instead of molecules, softspaces admit certain kinds of information and prevent other kinds from gaining entry. This process is made easier using micros programmable by individual owners. Essentially, the membrane scans the incoming signals for keyword matches or "hits" that the programmer has specified in advance, and rejects all others as junk signals. Note that this same method in the hands of the power-mad makes it possible to scan thousands of phone calls, letters, and other private communications for tidbits of political data.

1.7.2 Filters

There is no such thing as an immaculate perception. The signals coming into our nervous systems from the environment (light waves, sounds, temperatures and other skin stimuli) are partially organized and transformed even as they are sensed. Then, before these signals reach consciousness—if they do at all—they are further filtered and combined. We are not aware, for the most part, of which filters we have "in place" at any given time. If you have ever become conscious of your own ability to pick out individual conversations in a room full of partying people, you have direct knowledge of the processes involved, even if you don't know how they "work."

Our personal information environments, as manifested in our soft-

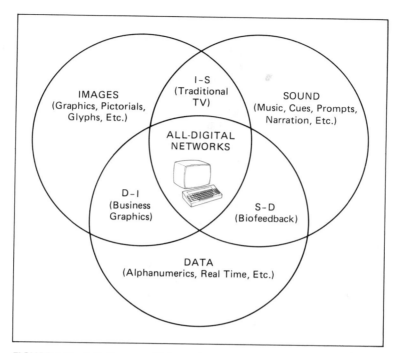

FIGURE 1.7. A Softspace Palette. In creating your softspaces, you currently have choices in each of the three major media arenas: images, sound and data. The bracketed examples are not exhaustive but meant to be suggestive. In the future, all-digital media will give the netweaver access to all three major forms of communication in a single tool.

spaces, will have certain integral, built-in filters as well as consciously programmed filters in place. For example, if your primary micro is an Apple IIe, your softspace is likely to have a lot of IIe information in it, and a lot less information about the TRS-80. The micro itself will "filter" programs intended for other machines, and will only run programs made specifically for that machine. Filtering is largely an unconscious, passive process, but it can be made more conscious as one becomes more aware of the information environment and its relevance to one's own needs and goals.

1.7.3   The Infosphere

The infosphere consists of both infotrash and infowealth. Which is which will depend on individual values, needs, and goals. What I filter out may be of great interest and use to you. Infoglut is the one malady that truly goes away when you ignore it. Information is like that. The infosphere, potentially, encompasses the entire planet. What *your* infosphere is, however, depends on you, your experience and development, your abilities and skills and capabilities, your lifework plan.

Your infosphere is composed of all your *sources of information.* If you are an info have, your infosphere is vast and for practical purposes unlimited. If you are an info have-not, then your life is probably dominated by mass perceptions transmitted to you from a limited number of information sources. NBC's "news" will also be your "news." It is not necessary to remain an info have-not. The barriers between have-not and have are being demolished as more and more people move from an Industrial Age mind-set to an Information Age mentality.

1.7.4   Media

We usually think of media as the *mass* media of radio, television, and newspapers. However, the emerging information environment consists of much more than these traditional media. The electronic media generally, including such new technologies as videotape and remotely accessible data bases, are becoming interchangeable from one form to another as digital technology and micros take over in all sectors of today's information economy, creating entirely new classes of "process" or "business."

This merging of media, resulting from the fact that information in any form can be translated into digital bits, has been called "immedia," "intermedia," "new media," and "micro media." By any name, as far as transmission networks and information systems are concerned, all media are the same. High-quality photographs only take 2 million bits to specify. A color television frame: 1 million. A typical interoffice memo takes only 3,000 bits to encode. Mixing different kinds of messages is as easy as mixing bits—almost trivial.

1.7.5   Content and Conduit

Diagram 1.7.5 classifies old and new media businesses, as well as other sectors of the information economy, according to their mix of content versus conduit. A conduit can also be viewed as a channel for information. Content is the information itself. Networks are conduits. Information is content. Sometimes the conduit or channel is also the message, as in "The medium is the message."

In the marketplace, information content—ideas, memes—is still relatively competitive. It is in the area of conduits—access to channels for content—that is relatively scarce and noncompetitive. Netweavers will help change that drastically over the next few years.

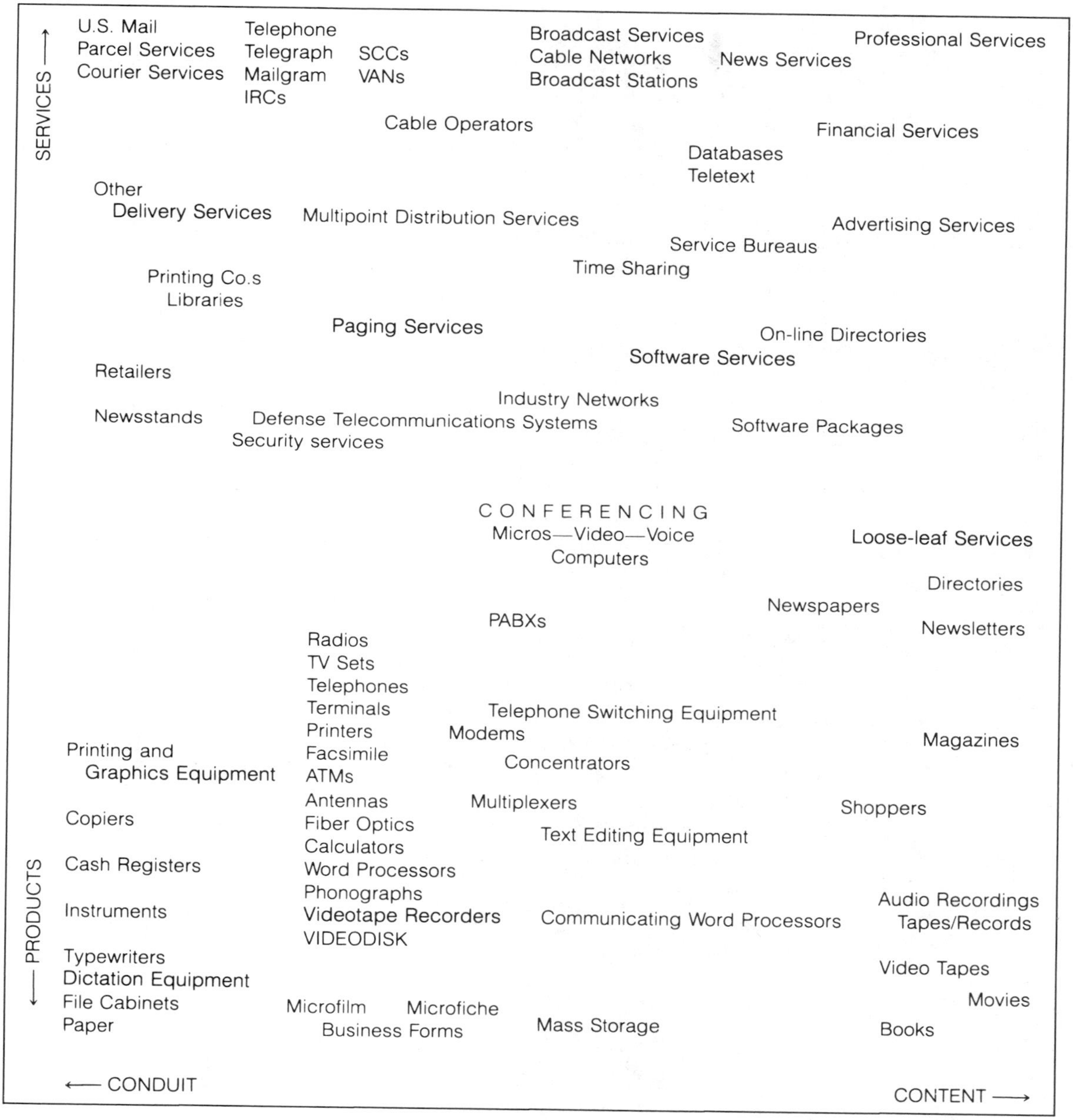

U.S. Mail   Telephone   Broadcast Services   Professional Services
Parcel Services   Telegraph   SCCs   Cable Networks   News Services
Courier Services   Mailgram   VANs   Broadcast Stations
                     IRCs
                               Cable Operators   Financial Services
                                                        Databases
                                                        Teletext
Other
  Delivery Services   Multipoint Distribution Services   Advertising Services
                                                   Service Bureaus
    Printing Co.s   Time Sharing
    Libraries
                       Paging Services   On-line Directories
                                                   Software Services
Retailers
                                   Industry Networks
  Newsstands   Defense Telecommunications Systems   Software Packages
              Security services

                              C O N F E R E N C I N G
                              Micros—Video—Voice   Loose-leaf Services
                                  Computers
                                                                      Directories
                                                   Newspapers
                              PABXs   Newsletters
                  Radios
                  TV Sets
                  Telephones
                  Terminals   Telephone Switching Equipment
                  Printers   Modems   Magazines
Printing and   Facsimile   Concentrators
  Graphics Equipment   ATMs
                  Antennas   Multiplexers   Shoppers
Copiers   Fiber Optics
                  Calculators   Text Editing Equipment
Cash Registers   Word Processors
                  Phonographs
Instruments   Videotape Recorders   Communicating Word Processors   Audio Recordings
                  VIDEODISK   Tapes/Records
Typewriters
Dictation Equipment   Video Tapes
File Cabinets   Microfilm   Microfiche   Movies
Paper   Business Forms   Mass Storage   Books

←— CONDUIT   CONTENT —→

SERVICES →

← PRODUCTS ↓

FIGURE 1.7.5. The Media

Adapted from U.S. House of Representatives Subcommittee of Telecommunications Report (full reference in Appendix 1.7.5).

_____ 1.8   NEWS

What's news? It all depends. On a personal level, in face-to-face contacts, news is gossip. Your news may or may not be the news of so-called news professionals. Some kinds of news are only gossip put to print or television signal. The gossip of some networks is the news of others. The mark of status for many journals and journalists is to be first with rumors that later turn out to be true to the facts, which they seldom are. The status associated with this feat, however, is so huge that people keep trying anyway.

The creation of news is easily done, and for those who know the rules, the news professionals are easily manipulated. In newspapers, for example, the news consists largely of classes of events happening over and over again with slight variations in specifics—a fire down the block, a car crash, a kidney transplant, a mugging, a murder, a launch of the space shuttle, et cetera, et cetera. For "professional news organizations" of all kinds, whether CNN (Cable News Network) or the _New York Times,_ news is the reproduction of the past. Never an accurate reflection of the present. Even less a guide to the future.

Since softspaces as they are evolving permit the production and manipulation of information as well as the gathering of information, the potential for self-definition of what constitutes news is huge. In the infosphere, each softspace generates its own news. How widely the particular news is disseminated is coming to depend more on the interests of the receivers of the news rather than the prestige or authority of its source or the size of its conduit measured in number of receivers.

News that is automatically disseminated to masses of people will, by definition, be of little value to any of them not equipped with the means to shape such common information to their own needs. Likewise, people who have the means to shape their own information will be less and less likely to use mass-based "news" as input for their own information processing.

1.8.1   Newsmaking

"Don't feed your mind with Big MacInfo Burgers." The making of news draws upon and reproduces organizational structures. There is no such thing as "objective" reporting, any more than there are immaculate perceptions. Identifying information sources and the kinds of information to be called "news" is part of newsmaking. It is helpful for the netweaver to ask of self and others, "What's news?"—that is, what kinds of information do I need and want to know? Since newsmaking is an active process of filtering and transmitting, validity checks need to be built in. Decisions made on the basis of a single source of "news" are more likely to go awry than decisions made after source-checking and verification processes. Part of the verification process should include an assessment of the various filters through which the news had to pass, including one's own. News and newsmaking are primarily social phenomena serving to create and reinforce the realities we hold as "true," "good," and "beautiful."

1.8.2   Intelligence

Intelligence, like love and consciousness, is quicksand territory, quivering with speculation, argument, definition, and counterdefinition. For netweaving purposes, though, concern is with two different senses or forms of "intelligence." The first kind of intelligence is a "quality of a system." It represents *potential* in one or more areas of human intellectual endeavor. IQ tests attempt to "measure" this quality in human beings. Artificial intelligence researchers take small steps in the direction of their goal primarily by chunking bits and pieces of human intelligence into forms that can be simulated (duplicated?) on information processing machines. An "expert system" is one that duplicates the decisionmaking and problem-solving ability of some human expert in a particular field, such as law or medicine.

The second sense in which netweaving is concerned with intelligence is in the receiving of information and/or news. To be "intelligent" is to be able to acquire and apply knowledge and information. Intelligence concerning a subject or proposed decision refers to the entire body of held information on the subject. From this point of view, the Bucky Fuller axiom that "we always know more, not less" is a hallmark of an intelligence system. Whether this system can apply what it knows is another matter altogether.

1.8.2.1 PERSONAL/ BUSINESS   Personal intelligence concerns knowledge about oneself and one's personal environment. The purpose of many human relations workshops and human potential training courses is to increase self-knowledge and self-awareness. Self-knowledge shades into awareness of relationships and process in relationships. Couple relationships (dyads), in which two people form a more or less stable system over a period of time, are subjects of intense analysis in American culture, both formally and informally. Most of what passes for "news" in popular yellow sheets is merely gossip about the latest twists and turns in dyads composed of the rich and famous, who seem to "network" only with each other.

The idea of "measured intelligence" using "instruments" such as IQ tests and related psychological testing can be safely ignored by practical netweavers. Useful, pragmatic definitions of individual intelligence include:

- Being able to work effectively at tasks involving abstractions;
- Being able to learn;
- Being able to embrace new situations and value from them.

Individual intelligence, in short, is a real-time, real-life quality brought to our day-to-day work.

Business intelligence concerns knowledge about the tasks and environment of organizations formed for specific purposes. If the business is a sole proprietorship or partnership, personal and business intelligence will overlap a great deal. Much of the work needing to be done in the information environment, now and in the future, is work that *can* be done profitably and efficiently by one or two people linked together in either a dyadic relationship that shares the same softspace or in a two-person network linking geographically separate softspaces.

1.8.2.2
GOVERNMENT

Many agencies of government gather intelligence on citizens and organizations. Those concerned with law enforcement, taxes, or foreign spies are only the tip of any modern government's information gathering activities. Unclassified government information is accessible, in theory, to all citizens. Using government sources in personal and business intelligence is a way of increasing your own infowealth and getting back some of what you've already paid for in tax dollars.

1.8.2.3
MILITARY

First and Second Wave military intelligence concerned such things as the movements and emplacement of troops, missiles, and ammunition dumps in foreign countries. Increasingly, the intelligence demanded by the military is the same as the intelligence demanded by private citizens. Therein lies the inforub. Microchips have applications not only in toys, but in guided missiles and particle beam weapons. It is becoming increasingly impossible to separate information of a "strategic" kind from information of the "necessary" kind. This tension has within it great potential for further implementation of the Big Brother network.

1.8.2.4
MACHINE

No machine will ever be intelligent enough [1.2.1].

1.9   NETWORKS

A loose definition of a network is "a collection of related people." There are many other definitions and models for networks, depending on which most recently self-proclaimed networking expert you happen to check. Which one(s) you use in your netweaving activities may depend entirely on your personal taste, and more than one definition may be helpful, in fact. The term has been used so freely of late that everything from a journal and its readership to a society for professional engineers has been called a network. The number of small, intermediate, and large companies using the word in their company names is growing by the month. Although many use the term "network" in their names, not all are truly network-structured in their organization or operations.

Theodore Roszak, in his book *Person/Planet,* said that a network is "a loosely structured awareness of grievances and concerns among autonomous, situational groups."

Other writers have made distinctions between "formal" and "informal" networks. The problem with a formalized network is that it may lose, insofar as it is "formalized," one of the main advantages of a network structure: flexibility. Where connections are allowed to be made and

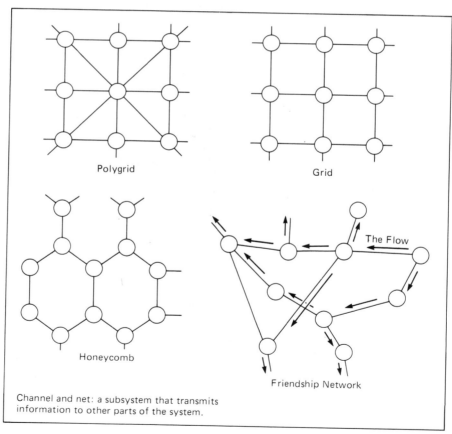

Polygrid

Grid

Honeycomb

The Flow

Friendship Network

Channel and net: a subsystem that transmits information to other parts of the system.

FIGURE 1.9. Some Network Structures

broken according to needs as individuals see them, networks grow and prosper. The best way to prevent the growth of networks is to attempt to bureaucratize, centralize, and formalize them. Even better, insist upon one definition of "network" to the exclusion of all the others that may be equally useful. The goal is to reweave and restructure entrenched institutions and professions, not to institutionalize and professionalize networks and networking. Un-networked (ugh!) structures will be at an increasing evolutionary disadvantage as more flexible networks organize themselves and literally overtake some of our most pressing systemic organizational and global problems.

Some have found it useful, for example, to make distinctions between "macro system networks" and "micro system networks." Or a similar distinction between "metanetworks" and "networks." If the network is open and flexible, nonhierarchical and distributed, such binary distinctions may not be very useful. No other network can get "above" the collection of "micronetworks" in order to "centralize" or "coordinate" the networks.

Another frequently made assertion is that networks, per se, are automatically nonhierarchical, more democratic and participatory than traditional bureaucracies, that they are somehow automatically perfectly flat, socially, economically, or—most important—politically. It may be that one distinguishing characteristic of networks is a motivation, on the part of its participants, to communicate across formal political and social boundaries. But I doubt if perfectly flat networks will ever appear. Whether they even *should* is also open to debate. In the Information Age, the importance of individual insight, ability, creativity, flexibility, intelligence, managerial abilities, and leadership should not be underestimated. These human qualities are not perfectly distributed among us.

Furthermore, to the extent that networks are still designed (and as long as computer and telecommunications technologies remain multipurpose tools), some people will continue to design and try to use hierarchical, centralized, mass-based networks. Some will want "global reach," sometimes just for the thrill that reaching large numbers of susceptible consciousnesses can give them. That will become less and less possible as people discover how to filter out mass transmissions that are of no relevance to them. In the long run, it may be that the more extremely hierarchical designs will die out because of their inherent disadvantages. On the other hand, it may also be that there are human beings on the planet who would actually prefer to live under something approaching a totalitarian electronic state simply to avoid the anxieties associated with freedom.

What is emerging out of all this is that a diagram of all the existing networks on the planet today would look like a knotted fishnet, with multitudes of nodes of varying sizes. Each node is linked with multiple others. The kinds and qualities and numbers of these links is changing over time, so that a static diagram can only capture a "snapshot" of the living system as it stands in a single point in time. If we could somehow model and display this master chart of networks, it would twinkle and blink and pulse with patterns and needs perhaps wholly its own. Trying to "meta" *this* network—in which we all exist—is literally to "play" the part of the All-Seeing I.

Does that mean that a netweaver should never design a "formal" network? Not necessarily. Just remember that your model must be subjected to frequent updating as real-world experience shows you the direction(s) in which the network is going. Later on in this sourcebook there are some techniques you might employ in "formalizing" and modeling your networks as they currently exist and then as you would like them to be. The kind of networks you participate in and invent will depend almost entirely on how evolved you and your current networks are. It may be that your network encourages the participation of all, that all voices are heard, that cooperation is valued over competition and that trust reigns supreme. It may be that your future networks will consist of totally free information flowing between perfectly free human beings (absolute freedom of movement, in physical space, and of communication, in information space). On the other hand, perhaps not. Much is a matter of design combined with intent. In fact, for the netweaver, intention may be everything.

In a bureaucracy, if you destroy the head, the organization is effectively crippled until a new one can be installed. The U.S. Constitution provides for quick and orderly substitution of heads in order to overcome this property of hierarchy. In a network-based organization, if you remove a node, the net as a whole continues to function. This is the rationale for the installation of military command networks. If control is distributed—whether of information, machines, or people—and made somewhat redundant in the network, it makes it very difficult for anyone to disrupt the activities of the network. This also makes nonhierarchical networks frustrating territory for "control junkies," who have strong drives to control everything, and a kind of negative paradise for paranoids worried about conspiracies. There are always new networks being formed, always new conspiracies. Infiltration of networks is ridiculously easy. But infiltration with intent to suppress or disrupt is much more difficult.

In traditional hierarchical industrial First and Second Wave organizations, complexities are simplified to the point of irrelevance. The interrelationships between problems and concepts, between problems and organization, are "organizationally unconscious," so to speak. It is highly desirable to seek out and perceive these interrelationships in order to make good decisions. The network structure, backed up by the appropriate conceptual tools, enables new and complex problems that are constantly emerging to be mastered. A *network strategy* will catalyze the appearance, development, and adaptation of interorganizational networks. These new relationships and systems should be more capable of dealing with the entanglement of problems at all levels of modern social systems.

1.9.1 Network Strategy Axioms

The following are eight axioms that apply to networks.

- All networks constitute and are constituted by softspaces (information systems).
- A softspace is composed of person(s), data, hardware, software for the person(s) (also called *selfware*), and software for the hardware.
- Networks are also composed of people, data, hardware (including but not limited to at least one microprocessor), and software, but, in addition, are defined by physical data communications channels between the various pieces of hardware and metaphysical connections between and among softspaces.
- Networks can be classified, for purposes of organizing and understanding them, according to the following categories:

  Global and transnational networks,
  Transcontinental networks,
  Broadcast networks (cable, satellite-print, television, radio),
  Special-interest networks (SPINs),
  Local area networks (LANs),
  Personal networks.

- Insofar as a given network is unique and ideosyncratic, it is or can be made to be proprietary. It can thus be marketed, as is any proprietary commodity. The content and the conduit aspects of the network, as well as its notable constituent softspaces, are the items of exchange which may or may not translate into dollars and cents, depending on how self-sufficient the given network has been designed to be. Expect, therefore, to see networks marketed as commodities in themselves.
- The value of a network lies not in how much information it contains, how many processes it controls or how much wealth it commands, but in the quality of the human beings who make it a living system.
- A group that lacks any form of data communications between its constituent softspaces is not yet a network as far as the netweaver is concerned.
- An "egalitarian network" is one in which any micro in the network, regardless of where it is located, can access all the other elements in the network, whether the other elements are micros themselves, or disk drives, printers, etc. This definition assumes that a given micro

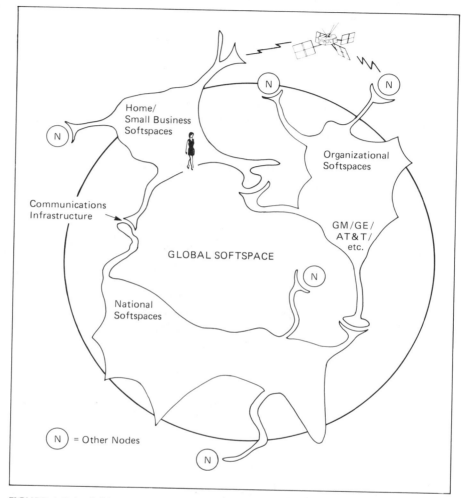

FIGURE 1.9.1. A Network of Networks

can be either "host" or "remote terminal" interchangeably with other micros in the network.

1.9.2 Networking

A brief collection of definitions of "networking" is as scattered as those for "network." It has been called "the familiar human activity of making connections with others to achieve some goal." By this definition, you can hitchhike across the country, getting names of people to mooch from in the next city from the people you mooched from in the last. Keep a diary of your exploits and—*voilà!*—you're a networker. You "work the net" for physical support, freebies, good vibes, and money—all quite *spiritual* at that. Variations on this theme abound. Calling networking "an exchange of information between like-minded persons" allows Jerry Rubin to turn social climbing into a New Age networking rage at Studio 54.

Networking certainly may *begin* with an analysis of the netweaver's current affiliations. (See [2.6] for more on how to go about such an analysis.) Networking can be carried out at a cocktail party, through the exchange of personal or professional data. It can also be carried on via telecommunications channels, and often is.

Networking can stop at networking, and usually does. To go the next step, to create a network where none existed before, requires more.

1.9.3   Netweaving

The difference between a networker and a netweaver is similar to that between a traveler and a tour guide. Or a hitchhiker and a highway designer, to use a more extreme metaphor. A netweaver knows how to travel—as driver or hitchhiker—but the goal of creating new networks is foremost. A netweaver must begin by becoming more conscious of the possibilities for networks in his or her current social sphere. Then, new connection possibilities for others may emerge. If the netweaver's current informal affiliations are purely local, based on face-to-face contacts and occasional voice phone calls, then it is not wise to try building a transcontinental or global telecommunications-based network tomorrow. Development of one's personal softspace, and the application of the appropriate micro technology to that softspace, is the place to begin. This then becomes the "seed" for all further netweaving. Think of it as a growing organic thing. You can't make cuttings for your friends, or transplant any part of the network until it has been nurtured to sufficient size and strength in your local greenhouse.

1.9.4   Network Organization Development

Beyond communities of consciousness and special interests, networks may be formed to achieve some specific goal, physical or social, or both. An example of a physical goal would be a global network aimed at the building of a planetary spaceport located near the equator. Once its purpose has been established, knowledge of how to ground and stabilize a network long enough for the job to get done will be essential. This is the province of Network Organization Development (NOD). Knowing how to create self-stabilizing human organizations is essential in the Information Age. Knowledge about how to create and maintain a "task force," "study group," or special-interest "commission" is becoming more widespread.

1.9.5   Dissemination and Utilization

A big problem with current scientific advance has been identified as the problem of "dissemination and utilization of scientific knowledge." This might also be the problem of "culture lag." It also contributes to infoglut. Certainly this is true in the area of telecommunications and the new media. For even though our contemporary sense of things is that "it's all moving too fast," in certain ways, and with certain seemingly intractable problems, things aren't moving fast enough.

Take, for example, the large system problem of acid rain. Many individual factories and power plants contribute to the atmospheric system a portion of the sulfurs and other chemicals that produce rains that kill the lakes and trees on which they fall. Individual factories or power stations can claim they're not responsible, since *their* toxic emissions cannot be "scientifically proven" to be a part of the problem. There are no feedback loops in place to control systems of this magnitude. What loops there are, are too weak or too slow to be effective.

There are thousands of insights, ideas, approaches, exercises, techniques, methods, and findings from scientific research just waiting to be fully tapped by the appropriate systems and subsystems of our society. These come from physics, sociology, psychology (esoteric and otherwise), space research, communications research, operations research, and medicine, to name a handful. Findings in biology, neurology, epidemiology, and physiology are coming in every day at a rapid rate.

Whatever one's definition of "computopia," nothing remotely approaching such a state can be achieved unless we learn to apply the wisdom we already have, and rapidly.

In this arena alone there are definite advantages to the network strategy. Past research has shown that properly designed and implemented network structures can incorporate the following characteristics:

- A tendency to encourage the full utilization of innovation at the individual and group levels, where innovation can have the most impact;
- A tendency to minimize the consequences of failure, on both social and individual levels;
- A tendency to promote the penetration of memes (ideas, artifacts, algorithms, methods, etc.) across socioeconomic barriers, while preserving ethnic and vernacular values: diversity-in-unity, autonomous participation;
- A tendency to maintain flexibility and adaptability in the face of new situations and developments;
- A tendency to put a high value on egalitarian rather than authoritarian roles and relationships.

Thus, combining such networks with the problem of dissemination and utilization of scientific knowledge will help our institutional systems and subsystems get smarter faster. This is probably essential to the survival of our species. Electronic computerized networks are tools for intelligence increase on a vast yet personalized scale.

1.10   SYNCHRONICITY AND GENERAL SYSTEMANTICS

You never know until later.
Systems exhibit systems behavior.

If it functions as expected—be suspicious. Carl Jung was a wonderful scientist. He maintained that just because data were irrational, that did not mean that they should be dismissed.

If groups possess "group consciousness," "group norms," and the like, and if organizations and societies are, in fact, large-scale living systems, then we should expect to be able to discern patterns of thought and behavior in these systems. We should expect to experience instances in which two or more apparently unrelated events become connected in and through our individual consciousness. We should expect that patterns will be formed in many different ways in living systems. Thus you may be thinking of a friend and get a phone call from that friend within a few minutes of the thought. Or you read an odd item in a book and find, in that day's mail, another reference to the same odd item. You get to work, and two people at the water cooler are talking about the same silly thing.

Synchronicity. A meaningful coincidence.

Synchronicities abound in the information environment. Three Mile Island merges in mediaspace with *The China Syndrome*. Who planned it?

As netweavers begin to deal comfortably with more channels of information and a greater amount of personal information processing on the contents of the varying channels, they will begin to experience higher rates of synchronicity. Seemingly separate channels will begin to correlate in highly specific and uncanny ways. It is recommended that netweavers keep a log of the synchronistic info-events they experience.

When one first begins to experience synchronistic phenomena, it can be somewhat disorienting and create anxiety and/or disbelief. It can also lead to paranoia. (Someone, Somewhere must be creating these coincidences, therefore I am being manipulated.) There probably is no Cosmic Coincidence Control Center. But in case there is: you've been warned.

If the telephone rings today: water it!

—*Principia Discordia*

Synchronicities may be the results of the synergies of whole systems humanity has not yet identified nor comprehended.

Synergy. A behavior or manifestation of a system that cannot be predicted by examining or understanding the behavior of its parts.

Complex systems manifest unexpected, not to mention unpredictable, behaviors.

Garbage In does not always equal Garbage Out. Great artists regularly turn sows' ears into the latest softwear.

Some garbage never does get collected.

Old Systems = Old Problems. New Systems = New Problems.

A systemist observation: The qualifications needed to get elected to office do not include the ability to function once elected.

Systemism. The state of mindless belief in systems; the belief that systems can be made to function to achieve desired goals. The state of being immersed in systems; the state of being a systems-person.

2

Personal Systems

Each of us carries around a 13- to 30-billion-cell biocomputer (depending on whose estimates you read, but who's counting?) inside our skulls. Our personal information-handling capacity evolves over our lifetimes. Within the brain's walnutlike topography, biochemical labs make drug-molecules that serve as messenger-signals around the system. We make, within our biocomputers, our own opiatelike drugs in response to stress and pain.

The construction of your brain was carried out under the direction of information coded in the DNA molecule—the chemical template of all terrestrial life. Each of us comes into the world as unknown and largely, but not completely, unknowing. Various wired-in programs get us off to a running start on learning. These programs have to do primarily with individual and species survival.

Brain/mind research over the last decade or two has made tremendous strides in discovering and decoding the various circuits of the brain. Early working maps and operations manuals for the human biocomputer are being drawn by neurophysiologists, biochemists, neuroanatomists, psychologists, hypnotists, information and systems theorists, cyberneticians, applied behavioral scientists, and physicists—all together in glorious interdisciplinary harmony.

How can a netweaver benefit from this explosion in generalizable self-knowledge? Does it do you any good to know how your own biocomputer works, alone and linked with others in networks? In a word, emphatically, yes! An information processing model of the brain/mind system is useful in understanding the workings of the self and in the design of your networks. In fact, just such a model has been used in psychology for quite some time.

This section will help the budding netweaver to perceive, understand, communicate with, and control (through goal setting and planning) his or her own biocomputer. In addition, the brain-as-model is useful in understanding and building networks of all kinds. The understanding of telecommunications on a global level is made easier if one understands the pragmatics of communications on the personal level. Without that kind of understanding, netweaving is a touch-and-go proposition, open to manipulation by others whose motives might not coincide with your own. Individuals must be both "DNA" designers and "neuron" participants in the new networks.

2.1 SELF AND BRAIN

All of what human beings know about the universe—all science, all technology, all psychology—has come through the brain/mind system. Understanding how this system works, and how it affects the study of all things we consider, including telecommunications networks and the creation of new, special-purpose networks, is crucial to our collective unfolding future.

Mastery of the new—new vocabulary, new technology, new skills— implies some ability on the part of the individual to reprogram and install new programs in the personal biocomputer. Successful survival strategies in the Information Age will depend on the fact that human beings can learn and adapt very quickly. In fact, more quickly than the groups and organizations in which they find themselves. The evolutionary advantage in the Information Age is weighted heavily in favor of individuals and organizational forms that respond quickly to new information and feedback.

2.1.1 Neurons

Neurons—nerve cells—are the basic building blocks of nervous systems. There are many different types of neurons in the catalog of neuroanatomy. Each is capable of responding to the inputs from other neurons connected through its *axons* and *dendrites.* The connection between nerve cells isn't really a connection in the same sense as wired mechanical components are "connected." The nerve cell connection is called a synapse. In response to signals within the neuron, the *bouton* at the end of one nerve cell releases chemical messengers—neurotransmitters—that cross the *synaptic cleft* and trigger a response in the connected cell. Each cell can have multitudes of connections.

Sometimes, inputs to the cell inhibit a response in that cell rather than trigger one. The speed with which the biocomputer works is a function of both electrical and chemical/physical processes.

New connections are being made all through the lifetime of a particular nervous system. Old connections may be broken. It is even rumored that a few neurons die every day.

The human brain and spinal cord (the central nervous system, or CNS—your personal biocomputer) is a vast internal communications network circulating electrical and chemical messages encoded with information you need to survive and self-actualize. The speed of the electrical impulses in the nervous system varies with the thickness of the nerve channel carrying them. In the smallest nerve fibers, impulses travel at fractions of millimeters per second. In the largest, signals travel up to 120 meters per second.

When you stub a toe, the pain impulse travels from toe to brain and back again in less than one-twentieth of a second. In terms of a single nerve cell, the cycle rate—the number of times the cell can be activated—has an upper limit of about 100 cycles per second. In contrast, electronic parts in computers have cycle rates of millions of activations per second.

It is useful to believe that the "average" human being, *given enough time,* can learn to do anything. It is assumed that the same will be true of the general purpose micro of the future.

The individual can function as a crucial nerve cell in Gaia's brain. Connected yet autonomous, the insight of the uniquely private individual can make important contributions to global thought patterns.

2.1.2   Brain Centers

Diagram 2.1.2 shows an idealized neuron. This cellular unit is the emblem of the netweaver. With concepts from information and cybernetic theory superimposed on its processes, the neuron becomes a model for the building of the world brain. Localized collections of neurons, and brain subsystems composed of distributed neurons, constitute *brain centers.* Subsystems for hearing, sight, taste, smell, balance, language, memory, and so on, have been mapped by brain scientists.

Each of us lives in the universe of our brain centers. We can only sample the complex energy states that surround our bodies: light, heat, pressure, and chemical compositions. We construct everything else from these samples, by inference.

At the level of our sensations, your images and my images are pretty much the same, and we can readily identify them to each other by verbal descriptions or some other common reaction. But as we update our neural maps of the world around us, we combine each image with genetic and stored experiential information. That combination is uniquely private, each to each. The higher levels of perceptual experience are personal views from within.

As telecommunications networks are constructed, clusters of individuals will begin to form interest and processing centers, devoted to

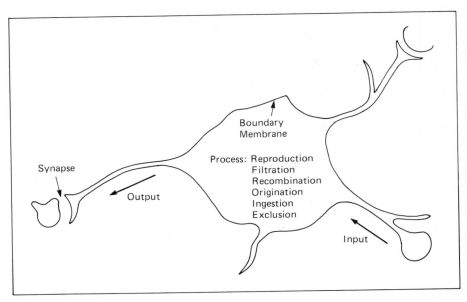

FIGURE 2.1.2. The Ideal Neuron

particular aspects of personal and social functioning. These world brain centers need not be large to have an effect on the systems and sub-systems of the planet. Elementary cybernetics tells us that small inputs can have large effects. Conversely, small inputs can inhibit potentially catastrophic reactions in a given system.

> No special interest or specific need is "insignificant" in the Information Age.
>
> The netweaver must recognize *needs* as being significant messages within human living systems.

2.1.3   Inputs

There are about 2 million visual inputs to the human brain. There are a "mere" 100,000 acoustic inputs. Most of us have learning styles geared toward our own sensory biases. Some learn best by listening to lectures and taking notes. Some learn best by reading (visual input, verbal) in a quiet corner by themselves. Some learn best by seeing a demonstration or pictures (visual input, graphic), combined with auditory explanation. It is useful to have a clear idea of your personal learning style (also called "cognitive style" by educators). Such self-knowledge can help you in choosing microcomputer hardware and software and can also help you in designing your personal communications networks.

This Information Resource Assessment profile checklist can assist you in charting your personal information and learning styles. Later on, you can use this assessment to chart your course through the coming decade of Information Age transformations.

Media assessment: List the magazines, newspapers, journals you subscribe to or otherwise read regularly or irregularly. List the programs you watch on television. List radio stations and content you listen to. Time spent in each activity should also be listed. If you don't know for sure, begin a log and keep it for a week. For now, *estimate.*

Marketplace: List the market trends you track. As a *consumer,* you need to know about the products you're considering buying. As a *producer,* you need to know about sources of goods and services to which you will *add value* in your work. Vendors, similar businesses or professions, fieldwork (visiting stores or calling the competition), trade shows, and trade magazines you use as sources should be listed. If there are government sources of info on your activity or marketplace, list them here.

Junk mail: List here the types and frequencies of the kinds of junk mail you are apt to save or notice. If you keep files of advertisements and announcements, note the names of the files here.

Books: List the books you have read (and can remember) in the last three to six months.

Software: If you already have a particular micro and a collection of software, inventory the software and add to this assessment.

Bookstores and newsstands: List here the names of the bookstores and newsstands you browse. List the kinds of information you are likely to look for when there. If you don't do this regularly already, make a note in your time/money planner to do so soon, and schedule regular visits henceforth.

Internal sources: List here your sources of personal information— reference library at home or at your company, files and file names, and so on. If your library is not cataloged, make a note now to work on this later.

Outside sources: List here your direct sources of outside information. These can be businesses, people, or institutions (libraries, for example) or other sources specifically oriented toward providing you with information, not goods or other services. If you already access some of the available on-line data bases, or a community network or three, list them here, too.

Personal networks: You are the center, the hub, of all your networks. List those in which you participate: trade associations, professional organizations, clubs, special-interest groups, political action committees, and so on. Be exhaustive. Diagram your personal friendship and business network(s).

You must know what your current inputs are, and how you currently deal with what you have, before you can proceed to the step of adding more to what you have. If you think of yourself as an "information have" from the very start, beginning with your unique profile of sources, interests and learning style, you can begin working with that info capital.

> There are many sources of information. In the Information Age, sources will multiply like tribbles. Advertisements will present given sources as being "all you need" and "comprehensive." This is not only presumptuous, insulting, and premeditated hype, but simply untrue. Before you run off to sample the latest data bank or "information utility," and especially before you plunk your money on the line for a given service, make sure you know what you need. Don't let someone else tell you what you need, no matter what their connections, corporate or otherwise. It is entirely possible that you won't need to go "on-line with the world" once you do your self-analysis.

You must also have a clear idea about what you can bring to your netweaving activities in the way of life experience and current information. The self-analysis you do will be invaluable in guiding newcomers to your network.

2.1.4   Filters

To date, no neuroscientist really knows *how* our brains store information. There are many elegant theories in the vying for top dog as Brain Model of the Week. Some suggest that we store information in our memories holographically and continuously, putting everything in storage and letting only a small amount seep through to consciousness. Others say that's not the way it is at all. Rather, our sense organs begin the process of organizing and filtering out signals before they ever reach the brain, and in the brain even more are filtered out before they are allowed through the short-term memory pass and into long-term memory.
     Most of those who study the brain agree that we do filter out a great

deal of what our senses would otherwise be able to present to us. We filter on the basis of interests, personal values, survival, drives, and needs. Further, we actively process only a tiny fraction of the remaining information (that which is still there after all our filters have done their work). Much of what we think, feel, and act upon is under the control of programs built in from the beginning and/or long-standing programs set on automatic.

In your netweaving and telecommunications activities, you'll be interacting with a wide variety of people and institutions, each with its own set of values and prejudices and goals. The situations a netweaver must meet will require thought, opinionmaking, and sharing. Decisionmaking—action—will occur in situations that are sometimes familiar, sometimes completely novel, as when interacting in an on-line community. Everything the netweaver does will be based on what gets through the personal filters of belief, attitudes, and values, whether those of self or others.

Netweaving will affect politics, aging, death, friendships, religion, work, leisure time, health, education, family, race relations, war and peace, rules and authority, material possessions, and personal tastes. Throw in art, music, and literature, and you can begin to appreciate the far-reaching effects of your activities as a networker, netweaver, and communicator.

Because personal values sometimes conflict with social values, and because the arena of personal and social choices is growing larger every day, it is important to reduce confusion on the personal level concerning one's values. *Knowing what you are likely to filter out is as important as knowing what you let into your mental map of the world.* It is also necessary to be able to decide consciously what you want to eliminate from your information input stream and what you want to add.

Here is a checklist to help you begin the on-going process of coming to know what filters you are using on your information stream.

- What beliefs and behaviors do I prize or cherish above all others?
- What beliefs and behaviors am I willing to publicly acknowledge, affirm, and defend, if necessary?
- What alternatives are there to my currently held beliefs?
- What are the consequences of each of my particular value(s)?
- Am I freely choosing each particular value, or am I being subjected to peer pressure or economic or political pressure to hold this value?
- Do I act my value(s), or just give lip service?
- Do I demonstrate consistency in my values, or do they contradict one another? Do I experience internal conflict in certain situations as a result of my contradictory values or prejudices?

2.1.5   Self-Communication and Creativity

The design process begins with you. If you do not experience good communications with yourself, it is not likely that your communications with others will be very good. The dictum "Know thyself" has never been more important than now, in the dawn of the Information Age.

Every individual, no matter what his or her success level may be, is looking to "make things better." "Things" in this context usually means the alleviation of a personal problem, an increase in economic or creative freedom, or the pursuit of a personal goal that may also have wider, humanitarian consequences. The big attraction of personal micros, networks, and telecommunications is that the humane application of these new technologies promises to "make things better" for the individual and society at large. Whether, in fact, this promise will be met depends mostly on you, how you perceive and implement your networks. The technology is next to useless in and of itself. Only when we link it with our own goals and requirements can we "make things better" using the design process and the technical tools at hand.

Creativity is necessary in the design of oneself. Creativity is a combination of feeling and knowing. All the technological jargon in the world, all the information in all the on-line services in the world, will do the netweaver no good at all without the wholeness that comes from knowledge combined with human emotion and feeling. To engage in the design of networks is to embark on a personal journey of design and the development of more creative personal behavior patterns.

It is easy to list, here, the behavioral "attributes of creativity" that netweavers need. However, in order to develop and refine these characteristics, the netweaver needs to know from the outset that social systems—the groups and organizations in which the netweaver must act—will make self-communication and self-development difficult at first.

The same social systems that welcome and readily assimilate the "creative product" may chastise and repress those activities required for such production because of its "nontypical" nature.

Netweavers take note: If your human relations remain on a "smooth course" during the process of your design activities, it is highly probable that one of two alternatives will hold true:

1. You are not being very creative.
2. You are working with people who are used to the "abnormal" behavior associated with innovation and creativity.

Fortunately, the requirements for good self-communication mesh with the required attributes for creative behavior in design:

- Freedom from pride, and therefore . . . ;
- The ability to be honest with oneself;
- Belief in one's own potential and ability to succeed;
- Freedom from fear of failure;
- Constructive discontent;
- The ability to escape from daily habit.

2.1.6   Memory and Prostheses

The memory storage limits of the human biocomputer are unknown. Obviously, its capacity is very large compared with any existing mechanical or electromechanical storage system for silicon-based computers. We are born with some programs "wired in." Those having to do with feeding, sex, avoidance of pain, for example, "come with the box."

Those of us not born with photographic memories—most of us—live a lifetime of finding ways to augment our memories. In the development of skills, such as driving or playing a piano, designing a network or soldering a cable, memory is developed through practice, practice, practice until the skill is "second nature" to us. For more abstract memory requirements, such as remembering our grocery needs or our bank balances or the RS-232C pin designations, we rely on prostheses—artificial devices designed to replace a missing part of the body or mind—to carry out our tasks: memos, notes, lists, wiring diagrams, floppy disks, and sourcebooks.

The most exciting aspect of personal information management opened up by micros and networks is this increase in memory capacity. This augmentation may lead to what one writer has called "the creative reintegration of the information society." On the other hand, it is also one of the most threatening aspects of the Information Age: the image of the Computer That Never Forgets.

2.2   PROCESSING: PROGRAMS AND METAPROGRAMS

What we *do* with the information that gets past our filters and into our world models contained in our biocomputers will determine whether or not such information truly is power. Most of the time, information is *not* power. The information we get in the daily news, for example, is largely in the realm of events over which we have no control and are largely powerless to do anything about.

What you do with information inside your personal information system once you have it is almost totally *process*. The process may or may not lead to a more or less tangible product. In biocomputer terms, processes are accomplished by programs and metaprograms of the brain. Programs are specific: do this, then do that. Metaprograms are more general, and control sets of specific programs. When you set a goal, you are attempting to install a metaprogram that will direct all the programs at your command toward the achievement of the given goal. Short-term goals are usually easier: go to dinner; write a report; call a friend. Long-term goals are harder: build a network; lose twenty pounds; go to the moon.

Your biocomputer, like your micro, is a general-purpose tool. The more programs you know how to run, the more flexible you'll be and the higher survival value your information system (you plus all your mental and physical extensions) will have. The installation of new programs and metaprograms is a lifetime process. Getting rid of older, less useful programs is also a lifetime process.

Self-actualization, self-development, creativity, and related "consciousness-building" concepts can be embodied in the following metaprograms.

Bring all existing programs and metaprograms to consciousness.

Continually add new programs and metaprograms.

Continually delete old programs and metaprograms that have become obsolete.

2.2.1   Lifetime Learning: Processes and Checklist

To decide and choose adequately, freely, one must have information. Enough information. You must determine how much is "enough." Adequate information facilitates planning and choice. In an age of accelerating innovation and change, the only learning/educational strategy that makes sense is a lifetime one. Scholarship must be diffused out of the traditional "centers of learning" and into personal and organizational softspaces.

Figure 2.2.1a shows the lifetime metacycle of information for systems from self to global.

One aim of your personal softspace design, including netweaving activities and system building for and with others, should be to enable you to acquire needed information quickly, economically and easily.

Raising your "information consciousness" means that you will begin to *think* in ways that will result in your being well-informed, more resourceful, and better able to metaprogram your own processes. This will cost you more in time than in money. Your human energy is possibly the most valuable asset you can apply in the Information Age. On the other hand, lack of information will cost you, too, as you overpay for goods

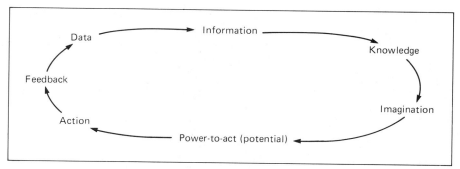

FIGURE 2.2.1a. The Life Cycle of Information

Zero	Information Content →	Enough
Guess		Decide
React		Choose
Failure		Success
Poverty		Wealth

FIGURE 2.2.1b. None versus Enough. As information content increases, so does the tendency of one's decisions to become better. Without adequate information, you are blind and impotent. With adequate information, the odds get tipped in favor of increased self-actualization.

and services, data and energy; as you waste your time and have many unpleasant experiences. The Information Age is an age in which those who know how to acquire, store, organize, access, distribute, and "control" information will be the power holders and the power brokers.

Data are facts, bits and pieces, without shape. *Information* is relevant to your particular viewpoint or frame of reference. You give data their relevance, their arrangement, coherence, and usefulness. *No one can "give" you information.* Your meaning, intent, and interest impart whatever information data can have.

An incomplete list of processes:

Reading
Watching
Listening
Interacting
Commanding
Analyzing
Anologizing
Responding
Scanning
Clipping

Filing
Juxtapositioning:
  collage
Connecting
Patterning
Dialectic
Footnoting
Highlighting
Excerpting
Writing
Paraphrasing

Rewriting
Note-taking
Message sending
Summarizing
Editing
Prioritizing
Rearranging
List making
Tapping the Great
  Unknown
Meditating

A personal or business information system should enable you to acquire, process, and store information quickly, economically and easily. Further, it should enable you to transmit the results (messages) to the right person(s) or organization(s), at the right place(s) and at the right time(s). These messages can take a variety of forms: software, reports, manuals, letters, memos, floppy disks, and so on.

People with "high information consciousness" generally exhibit the following characteristics:

- They are generalists in the sense of having a general systems orientation.
- They tend to read a lot.
- They tend to keep and maintain files on all kinds of things.
- They are curious about all the things, processes, and events of the world.
- They prefer their own primary sources and networks—informed firsthand opinion—for information of most kinds.

2.2.2   Goal Setting, Goal Seeking

All processes lead to end states. The end states are determined, in part, by the goals of a given system. Sometimes goals are reached by setting intermediate *objectives*. Systems that set and seek goals are called *purposive systems*. Philosophers are fond of debating whether or not the system called universe has a purpose. Only you and I can decide whether *we* have a purpose.

In personal terms, being concrete about one's goals and purpose(s) is a prerequisite for good decisionmaking in any realm. Write out your goals. Review them daily. If the goal is long-term (to learn how to write machine language programs), then breaking the goal into short-term objectives that can be accomplished in a given time will enable you to measure your progress (today I will learn three new machine language commands).

There may be some confusion concerning "objectives" in some systems with which you must work as a netweaver. Some of your unconscious programs and metaprograms may also generate such confusion on a personal level. The participants in a given system may be able to state what their objectives are. However, the statements they make may have a number of purposes that are independent of the actual performance of the system.

To cite an example, a university president needs to have as large a budget as possible for the university's operations. Consequently, she must appear before legislative committees and the public. The president states the "objectives" of the university in as attractive a manner as possible. "Quality of education," "faculty excellence," "public service," and similar phrases will be bandied about as "objectives."

Businesses also present glowing pictures of their objectives. This is done not only to attract customers, but also to attract satisfactory investment funds.

On the personal level, objectives should always be examined for hidden agendas and issues that may be lurking below the surface. One way to do this is to make sure that objectives are always presented to self and others in terms as concrete as possible. Not "self-actualization" or "better health" or "the good of the people," but specific goals, such as "jogging two miles a day" or "adding another computer to the network," or "reducing infant mortality by 50 percent."

2.2.3   Planning and Planning Aids

Planning usually involves at least two phases: the data-gathering or inventory phase and the action phase. Too many "action-oriented" people skip the data-gathering phase altogether, wanting immediately to "join a network" or "start a network." They jump to the action phase and start publicizing or talking about their new "network." Or they buy a subscription to one of the mass information services. They attempt to recruit people to their network without knowing much beyond the fact that they want to start one, and have something called a "network" to call their own. Having skipped the data-gathering and planning phase on the personal level, they then are unable to present themselves or their

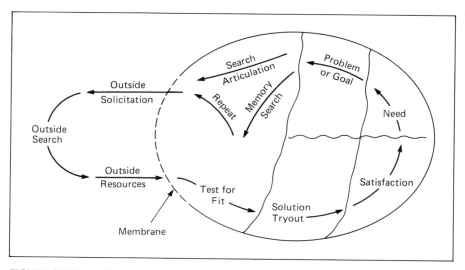

FIGURE 2.2.3. Individual Information Model

information resources or skills in ways that get much results. Consequently, technical and social problems soon mount up, and the vague plan is abandoned in a shambles of discouragement, gloom, doom, and blame.

Planning aids include computerized spreadsheets for budgeting, flowcharts, time-task charts, PERT and CPM network diagramming, and a host of others. Probably the most useful planning aid is the old-fashioned paper and pencil. Thinking is no substitute for action, but action isn't very effective at all without thinking.

2.2.4   Human Life-Styling: Design

A "life-style" is you and your personal system: what you eat, drink, breathe, think, and do. Life-styling as a process assumes that you take charge of designing yourself. Successful designs (as measured by your own health and sense of well-being or peace of mind) will require information about ecology (your personal environment), nutrition, exercise, and stress reduction.

2.2.5   Time Management

The achievement of goals and objectives requires time and energy. The popularity of books and advice on "one-minute management" and "thirty-second employment" attests to the widespread perception that time is not under our control. First, we have to face the fact that nothing is ever *completely* under our control, time least of all, since we are locked quite securely in the Eternal Now. But, once acknowledged, this grim reality can become our ally in the management of time.

There are literally hundreds of approaches to time management and task scheduling. Self-management and time management go hand in hand in the Information Age because more and more autonomy is being assumed by independent information workers.

The following metaprogramming checklist should be referred to during any planning process.

- Make sure your goal(s) are framed in Information Age concepts. If you have not done so already, using the concepts in this sourcebook, reframe your life and business in Information Age terms.
- Determine the overall goal and write it down.
- Break the goal into subtasks or objectives (sometimes also called milestones). These are all the little steps that must be taken to get to the main goal.
- Set a time goal for the overall project. This is the total amount of time you have at your disposal to get to your end state.

- Each of the subtasks should be allotted a percentage of the total time available.

- Make a graphic, written display of your time line: the overall time, the individual tasks, and the times allotted to each task.

- Throughout the project, you then use the master schedule to check your progress. If necessary, redo the schedule at set points in the project. Time estimates for some tasks may have been too generous or too miserly.

- A good rule of thumb for any task time estimate is to take the time you first think it will take and then double it for the master schedule.

- Above all, try to be realistic. Ambitious goals cannot be accomplished in a week. (This is probably the hardest part of managing one's own time.)

2.2.6   Creative Leisure and Stress Management

A rapidly growing sector of the information economy is that dealing with creative leisure. As automation frees up more and more human time, and as appropriate technologies enable us to do more with less, the total amount of leisure time available to society will increase. The old axiom that "an idle brain is the devil's workshop" may be even more true in the Information Age. If you suddenly had a work week that was reduced by half, what would you do with the remaining time? Many people, when confronted with this choice, have opted for second jobs. But what if there are very few, if any, attractive and ready-made second jobs to go around? Then you'll have to create your own second career.

Services providing alternatives to sedentary mass entertainment will continue to grow. We know that, even now, more free time than ever is being spent on physical fitness and health. In coming months and years, more attention will be placed on "education for leisure" and "leisure time education." This sort of education will provide individuals with methods and tools for identifying, comparing, and selecting leisure time alternatives.

From the earliest years, we need to be nurtured in ways of thinking, ways of feeling, and ways of coming to know. This kind of education can enable people to know themselves better—their talents, predispositions, temperaments. Such an education can enable even the least gifted to make maximum use of their gifts in satisfactory leisure time as well as work time. As writer Isaac Asimov has pointed out, if individuals can use their leisure time in self-satisfying ways, they most likely will also please others. We could then enjoy a world in which everybody finds everybody intensely fascinating and stimulating and entertaining. If a time traveler from 1984 were then to tell such a people that the general feeling, circa 1984, had been that in a leisure world everybody would loll around watching television and eating bonbons, decaying, these future ones might laugh. They might retort that the *real* decay existed long ago, in the 1984 world. In 1984, people were still compelled to do things they didn't like or want to do, nor had a talent for. How astonishing that, in 1984 still, the alternative was starvation. "It was odd," the time traveler would be told, "that you believed you had eliminated slavery, when you'd only just begun."

On the other side of the leisure coin is the workaholic syndrome. This has also been identified as "type A" behavior, leading to high blood pressure, cardiovascular disease, and other stress-related illnesses. (Some physicians have gone so far as to suggest that *all* illnesses are stress-induced.) If you are pursuing an independent career in some aspect of the information environment, there may be a great temptation to work much harder than you ever would have for any other employer. This is especially true in the early stages of self-management, when the economic pressures combine with the demands of the task(s) to make you believe that if you just work an extra five or six hours a day, you'll somehow "get caught up" and things will be all right.

Personal Systemantics Axioms: You'll never get caught up. Work expands to fill the time available.

It's important to learn how to recognize your own stress symptoms and learn how to deal with your own stress behaviors. This sort of self-management can take many forms: exercise, hobbies, meditation, bio-feedback, massage, and hundreds of others. The main thing is to actually make stress management a part of your daily routine, and to use your time management skills to incorporate periods of relaxation and fun into your life apart from the major goals you may have set in work. Even though work and play are merging more and more in the Information Age, cyclical breaks from what we are doing can help us bring fresh minds and healthy bodies back to the tasks at hand.

_____ 2.3   CONTELLIGENT SYSTEMS

Contelligence is consciousness combined with intelligence. The linking of human biocomputers with micros and networks defines a particular kind of evolving system. Since we have people in the linkups, we know that the system will display phenomena and states that we associate with consciousness.

In the arena of software design, programs for communicating and manipulating alternative world models and cosmological premises are lacking. Simple programs in this direction include things like a computerized *I Ching* or astrology programs. Programs for teaching elements of neurolinguistic programming (NLP), semantics, general systems theory, communications and information theory, and communications pragmatics are also lacking.

As they appear on the scene, the degree of interactivity of such programs might be one way to measure their level of sophistication and usefulness. Teaching games, self-help and self-analysis programs, or lifework planning programs might take the individual through a course of personal exploration of skills, interests, aptitudes, and aspirations. These program designs must combine the arts of the programmer, counselor, psychologist, and educator. Some of these kinds of programs might be best implemented on-line, where a live human being is standing by for human dialog and insight from another quarter, augmenting the programs.

2.3.1   Symbionics

Western medicine has emphasized the "magic bullet" approach to disease and health. Technology and the scientific method have given us sophisticated life support and life extension systems: organ transplants, kidney dialysis machines, artificial hearts, microprocessor-controlled limbs. These options have presented each of us with personal decisions and ethical questions. Shall I donate my parts to an organ bank? Shall I donate my ova or sperm to a fertility clinic? Should I make out a "living will" so that no "extraordinary" means are taken to prolong my life beyond what I consider to be worthwhile living?

As electronics and the life sciences meld, the number of such questions will multiply. Some experts are already thinking ahead to the time of the "symbionic" mind: the possibility that microprocessors may be implanted in our bodies and connected with our CNS to enhance our abilities or control otherwise malfunctioning physiological processes. The initial stages of these developments have shown great promise. Implants containing drugs that must be administered in measured amounts over long periods of time are already in use. Some of these implants are radio-linked to the physician's office. The doctor can remotely monitor the physical processes involved and send signals back to the implanted devices to alter the amount of drug administered. These basic capabilities are direct outgrowths of space science, cybernetics, and telecommunications.

The ability to create and participate in a network is only one aspect of what, in the long run, must become an integrated approach to health monitoring, health care, and so-called health-care delivery. As usual, these capacities have their nightmare sides. Who decides who will benefit from increased technical capability? Will telemonitoring techniques be used to track "criminals"? More than one law enforcement official has suggested the use of radio implants to track parolees fresh out of prison.

The term "symbionics" suggests a positive approach to these questions by emphasizing symbiosis between ourselves and our mechanical creations. Mutual support and harmony are possible, but not necessary, outcomes. Our networks need to focus on sifting out—with great care and sensitivity—those technological possibilities that are truly symbionic.

Above all, the networks we create must, themselves, be symbiotic in relation to any given participant. Deciding what is mutually beneficial and what is merely parasitic must be worked out case by case by each netweaver. Looking at the possibilities inherent in your work as a netweaver shouldn't be regarded as useless speculation. The most farfetched science fiction of the past has become current ethical debate. It helps to look ahead.

2.3.2   Biofeedback and Suggestology

Combining biofeedback with computer programs will lead to the next generation of "healing games." In late 1983, the first of such programs hit the market. The design incorporated a low-cost galvanic skin resistance (GSR) measuring device that plugs into a micro. Using the GSR measurement as its input, the micro program allows the user to play a simple game and learn how to control the skin response at the same time. GSR correlates with one's brain waves, so it is possible, though not strictly accurate—yet—to speak of "controlling the computer with brain waves."

Eventually, more sophisticated sensors will be used to control more complex activities: blood pressure, heart rate, body temperature at various points, breathing, alpha and other true brain wave inputs. The results on our collective state of health will be dramatic. The Great Killer Stress may meet its final match in microcomputer games.

The science of accelerating the learning process, using multiple sensory stimulation and findings from biofeedback research, has been called "suggestology" or "optimalearning." By combining sensory awareness, relaxation, guided imagery techniques, and other "full brain learning" methods, a netweaver can improve performance of any given learning/doing task.

2.3.3   States of Consciousness

The human brain has millions of cells and billions of connections. A micro has only a few tens of thousands of parts. Now and for some time to come, we can safely posit that the human biocomputer is about 10,000

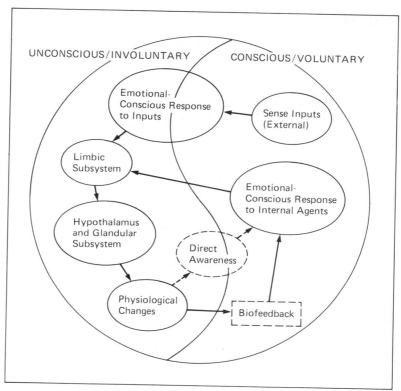

FIGURE 2.3.2. A Systems Model of Biofeedback

times more complex than our micro machines. Reprogramming ourselves is a more formidable task than reprogramming a machine. Fortunately, it is not an impossible task.

There are many esoteric religious and mystical traditions that have transmitted methods for altering consciousness. Their purposes range from magical to spiritual. It is likely that these methods for altering human consciousness can be augmented and accelerated in human/micro-computer combinations. If that turns out to be true, what kinds of possibilities are opening up in the age of micros and telecommunications? What kind of experiments are possible? What kinds of opportunities?

A lot of human learning is "state-dependent." That is, we learn more or less well depending on our state of mind or consciousness. Electronic intelligence amplifiers may serve us best as feedback devices enabling us to tune in to our best learning states. Rather than see micros as information shovels, loading us down with more and more infojunk, we might better use them as sense extenders, enabling us to get feedback from other living systems levels into our personal system.

2.3.4   Communities of Contelligence

Using still more sophisticated techniques such as evoked potential response averaging—a kind of brain wave monitoring—it may be possible to design contelligent systems that respond directly to specific human thoughts. Such systems would be useful in applications where the lag time between a human being's direct awareness of a situation, the conscious awareness of a situation, and an appropriate physical response is critical. They would also be useful for those who are physically unable to respond but are still fully conscious, thinking persons.

Communities of contelligence will form around tasks, long-range scientific and social goals, and special interests. To some extent, this process is under way even now. Large systems problems associated with space exploration have created a global contelligent community on the space frontier. This prototype community is a good model for netweavers to study, combining as it does both network-based and more traditional organizational principles.

2.4   TERRITORY AND INFORMATION

Human beings are territorial animals. Personal privacy and possession seem to be built into our biocomputers. Assertive realization of our territorial inclinations is an essential part of self-actualization. The effective functioning of human groups at all organizational levels depends on the degree to which territories are clearly defined. While the importance placed on territoriality varies across cultures, certainly those netweavers working in the United States, Canada, and western Europe will be able to help avoid and resolve many conflicts by acknowledging and working with the territorial behavior programs of self and others.

Many groups and organizations in society are trying to make information per se a "commodity" in the Information Age. Who gets to possess and who gets to disseminate various kinds and collections of information become essential territorial questions. Who owns the signals bounced off publicly financed communications satellites? Closer to home, does the individual have a right to determine to whom and how his or her name, address, and other life data are distributed?

Ownership is the essence of territoriality. It may be that, in the Information Age, we will all gain more territory overall if some territorialism is given up at the group and individual levels. *Territory* can cover a lot of mental ground, if you'll excuse the intentional pun. The object of ownership can be a stretch of land, physical objects, rights and privileges, ideas and information. Anything one seeks to own is territory. Who owns your network(s)?

In the Information Age, we arrive at an era when an individual or group can create new information territories at will. In this sense, there are no limits to the total amount of information territory available to the imaginative.

Territoriality is always expressed in personal terms, even though it may be implemented by large organizational resources or the force of law. For the individual, territory is that area of life experienced as one's own, over which he or she has control, takes initiative, or has expertise or for which he or she accepts responsibility. Territorial feelings and behaviors are closely linked to our sense of freedom. Privacy and territory are intimately related. The private domain is the territorial area staked out by an individual in order to secure personal privacy and security. This includes psychological spaces, physical spaces, and information.

A private space is a place where one can escape the attention of others. Much of this private space is maintained in our consciousness, in the form of information about ourselves and conversations with ourselves. We each have a right to defend this essential information insofar as it constitutes our private space.

2.4.1   Privacy

The right to privacy is a legitimate claim of individuals, supported globally by the doctrine of human rights, to determine for themselves when, how, and to what extent information about them is communicated to others. This claim can be extended to groups and organizations to a lesser degree, perhaps, but privacy, like territory, is primarily personal in nature.

Individual privacy can be maintained. It is neither necessary nor inevitable for electronic technologies to destroy privacy [3.3.4].

2.4.2   Ethics and Etiquette

Dear Ms. Telemanners:

Friends call my answering machine and then hang up without leaving a message at the sound of the beep. I am careful to leave my name and phone number on all the answering machines I encounter, but a lot of the owners never call back. Some of the answering machines I have called have been openly rude to me from the outset, sounding annoyed that I called at all. Why are people so rude to machines and with machines when they would never be so rude in person? And even my friends!

Alfalfa

Oh, Alfalfa, I know just what you mean. Now that all the groups and subgroups in our society are equal, the only things left to abuse are machines. Don't take it personally. What I mean is, don't identify with the machines, one way or another, and you'll be okay. Hang up on a machine yourself, sometime. You'll feel ever so much better for it. Remember the days when all you got on the other end of the line was a secretary telling you that Mr. Smith was in a meeting? You don't want to go back to that, do you? Besides, the function of answering machines is control over one's telephone communications, and that works both ways, for caller and called. If someone *really* wants to hear from you, they'll leave a message. Just remember, it's still perfectly socially acceptable to hate machines arbitrarily and capriciously, *especially* answering machines.

Dear Ms. Telemanners:

A friend of mine gave me a diskette for my micro with a copy of a best-selling program on it. I recently wired up my own cable TV descrambler from plans in *People's Electronics* magazine. Using my micro and instructions from other friends, I found out how to get in and out of MILNET, with computers all over the country to play in. I swap game programs with my friends all the time. And one of the community networks I use often has codes that let me decode copy-protected programs. A week ago, at about 3 A.M. on a Tuesday, two men in black pounded at my apartment door. They told me my apartment building was surrounded by the local police, the telephone security people, several large software company agents, Military Intelligence, cable corporation security agents, IRS enforcers, the CIA, the FBI, NSA, IBM, and a couple of Russian KGBers from the embassy down the street. I think there was at least one Japanese agent, too. They were all being coordinated by a computer security consultant. They waved guns at me and placed me under arrest. They confiscated all my diskettes and my micro and my modem, my descrambler and my collection of Playboy Channel videotapes. Now they tell me I am a computer criminal and bootlegger. I have been charged on three thousand counts of telecommunications piracy. I'm out of jail on a million-dollar bond, and awaiting trial. Can they do that?

Spanky

Spanky, honey, did you ask to see their identification or a search warrant? My advice to you is to get a good media lawyer from Philadelphia.

Dear Ms. Telemanners:

I wanted to build a mailing list for my new telecommunications consulting electronic cottage, so I traded a list of my friends for an equal number of names from other friends. After eliminating the duplicates, I found that I had nearly doubled my mailing list. I went through the same process again with a local computer club, and now my mailing list has over a thousand names. I'm going to sell the list to a large mailing list and marketing firm, making a tidy sum that I can throw back into my business. My question is, do you want to trade mailing lists?

Steveanne

Dear Steveanne, Why would I want to tell you so much about myself? After all, it's not as though we were in the same encounter group or something. Name swapping is a delicate matter and should be handled with great care. You should always get permission from the swapee. Do you want *your* name dropped any old place? Of course not. If I were your friend, I'd demand my name and address back.

Dear Ms. Telemanners:

I think someone is reading my electronic mail and has broken into my home micro. I have put up passwords and No Trespassing notices all over my personal information spaces, but I'm getting random graffitti in my financial files. I feel as though someone is continually looking over my shoulder when I'm at my keyboard anymore. Do you think it's the government? Am I paranoid or what?

Anton

Anton, sweetheart, I thought you'd never notice. I've had this crush on you for so long now, I had no other way of getting your attention. Ms. Telemanners used her secondhand Cray to tiptoe through your information domain. I didn't mean to frighten you. I hope you don't mind. How about getting together on that kinky system in New York this week? If the answer is yes, call me at this newspaper immediately! Let it ring once and then hang up and call again.

Dear Ms. Telemanners:

At home, it happens just as I'm sitting down to dinner. At the office, it happens just as I'm about to take a coffee break. I'm talking about those phone salespeople who just call up and start ranting without even asking if I can talk at the moment. Some of them have gotten quite good at getting past my secretary, who answers the phone at work most of the time. What can I do? It's driving me crazy. The number of these calls has increased significantly in the last several months. I don't need any more newspaper or magazine subscriptions, promotional calendars, or pens for my business, and certainly no more wholesale panty hose. Is there any way I can turn off the flood? Should I be rude to them, since they are being rude to me?

Clementina Atwater III

Dear Clementina, What you have been subjected to is *telemarketing,* and you might as well get used to it. If you are moderately wealthy, you can buy a screening device for your phone at home that will only allow the phone to ring when someone you know calls. You know it's someone you know because the someone has to enter a special number that you give out, *after* they have connected to your phone but *before* the phone has rung.

In the case of your secretary, you need to wise up this person so that the telemarketers employing various communicational "ruses" cannot get through to you. This requires knowing more about the breed. Mr. Alan Jordan is the author of *The Only Telemarketing Book You'll Ever Need* (Add-Effect Associates, Wayne, Pa.), and that's all I'm going to tell you. In the last resort, you can always be rude to them.

Dear Ms. Telemanners:

How can we get George to answer his electronic mail sooner? He doesn't look at it for days after it's been entered into the computer. He's slowing down our whole network. It wouldn't matter, except that we regard his input as vital to our network. Any suggestions?

Artemus Incliner

Dear Artemus, This is a complex question. I had an uncle once who let his mail pile up in front of the door for days before he'd get to it. Invariably there'd be a bill overdue or a grand prize announcement he'd missed because he opened the envelope after the contest deadline. He always seemed to get along. Had lots of friends.

When you say that George is "vital" to your network, what do you mean? Is he being paid to do something related to his e-mail? Or do you just send him notes to pick his brain for information that you might just as well get somewhere else? Is he a manager who must make decisions and answer questions based on his mail? Did he make an agreement to answer his mail instantly, just because it can be delivered instantly? Perhaps he's just being rude. Is e-mail *really* "mail," or just another intrusion by yet another electronic medium into our lives? Remember that you don't have to answer your phone or your mail if you don't want to, either. But if George is breaking his agreements with you, remind him of his responsibilities, in a polite way, naturally. Or maybe, like my uncle, George has died leaving quite a large pile behind him.

_____ 2.5 PERSONAL SOFTSPACES

Your softspaces will encompass both physical and informational territory over which you have some control and exercise competence. Your softspace is an extension of your central nervous system. More and more, your biocomputer will be enhanced by appliances you wear, such as wristwatches, pagers and beepers, and portable display terminals. Your main access softspace might be an office, den, or some other part of your living space. Links between the purely physical and the purely informational are increasing with every leap of technological capability. As a result, we are able to do more and more from just about anywhere.

As you add hardware [4.12], software [4.14], and links to your personal information processing system, your effective mental capacity will be increased. The degree to which you are able to take advantage of these new capabilities and tools will depend in large part on your imagination, your goal-setting and goal-seeking abilities, your communications and social skills, your ability to control your own time. Not only must fancy hardware be combined with software in a personally defined fashion, but your end uses must also be chosen well. If you're going to build a network, you'll have to know exactly what information territories you'll be claiming as your own, what territories you'll be expanding, and what territories you'll be willing to give up, trade, or delegate to someone else.

2.5.1 Hardstuff

Decision choices are always personal, even those made in corporate settings on behalf of an organization or group. Although they may be influenced by a variety of conscious and unconscious factors, the ideal is to make choices as consciously as possible. Today, microcomputer hardware is being sold on the basis of traditional appeals to status, economics, spurious comparisons, meaningless measurements of performance, love of glitter and flash, and all the rest of the Second Wave array of methods designed to whip up demand for mass-produced goods in a so-called competitive marketplace. Buying a computer on the basis of its nameplate is like buying designer jeans. Buying a computer on the basis of its cost can lead to the same disappointment that results from any purely cost-based economic decision.

Buying on the basis of group needs rather than individual needs can also be disastrous in the long run. "Compatibility" is often cited as the main reason for buying what the rest of the crowd has bought. But compatibility barriers are more easily overcome than barriers to full personal use that result when you buy a machine that doesn't have the software or peripherals you want, even though the rest of the group, company, or crowd is using the same machine.

The ultimate criterion for any given hardware system is its flexibility: how well, how fast, how easily can it be changed, added to, reconfigured to make it do the new thing I want it to do? Any machine that has only one input/output port, for example, is severely limited. As microprocessor-based machines get cheaper, it may be reasonable to dedicate a machine to this and another to that. In a softspace of one's own, though, that can be quite expensive in the short run.

Beware of equipment chauvinism! You are not what you own. Those who look down their noses at TRS-80s or IBMs or Commodores may be in the worst position to give you advice on anything, much less what to do with your micro or how to design your softspace. As a netweaver, you will encounter a wide variety of people and machines. They all can potentially contribute to the world brain. Every neuron is valuable. There is not, yet, a single universally applicable machine.

2.5.2   Softstuff

The conventional view says, "Select software first, then buy hardware." That view is largely a holdover from the mainframe, Second Wave era, when such a linear approach was more feasible. Systems designers today put together systems based on both hardware and software specifications. If the hardware is exactly right, they may even write new software for it rather than make do with a substandard machine. Hardware and software should be selected together, based on current and projected end uses. Even though a particular piece of required software may not be currently available on a given machine, the overall value of the machine and the other functions it *can* perform may still make it a great buy, even with the added cost of new programming. The conventional "software first" view implies that you only have one or two functions you'll want your softspace to perform. The really valuable softspace is able to do a wide variety of things, more or less well.

Good software combined with adequate hardware can help you expand your information territory. Bad software combined with excellent equipment won't take you as far. Often, what is "good" or "bad" for you must be determined on a trial-and-error basis. Admittedly, this can get expensive if you do absolutely no preliminary analysis, planning, or work to determine what you need, why you need it, and where you want to go with it. But some of what you invest must be counted as "learning," whether it be in the form of time or money. Allow yourself this learning leeway, especially where software is concerned, and especially in the beginning. You won't be sorry.

2.5.3   Ergonomics

Ergonomics originated with architects and industrial designers, especially in European schools of design. Ergonomics is the study of the physical human in the designed environment. It includes such things as optimal posture, work heights, seating, and control placement and reach. It is used in the design of automobiles and aircraft. It also studies things like eyestrain and the effects of equipment radiation emissions on individual health.

In the early 1980s, as video display terminals spread along with microcomputers, the National Institute for Occupational Safety and Health (NIOSH), prompted by concern from unions and consumer groups, did some preliminary study of emitted radiation from such things as VDTs and CRTs. (No, Virginia, cathode rays are not themselves x-rays, but they can *produce* x-rays, which are dangerous in prolonged exposure to human beings.) NIOSH found that employees using such devices have more health complaints than workers who do not use them. Complaints included eyestrain, headaches, backaches, and irritability. No doubt many of these complaints originated not in the radiation coming from the tubes but from bad seating, poor overall lighting, and improper working heights of keyboards.

There haven't been any long-term studies of the effects of even low-level radiation from video display devices, so the final word is not in yet on that particular issue. But it doesn't take much study to determine whether your workspace is ergonomically sound. You are either comfortable working or you are not.

From a stress point of view alone we know that no one, except under extraordinary circumstances (such as a harried author racing to meet a looming deadline), should spend more than four hours at a stretch in front of a terminal. Work should be combined with frequent breaks for exercise. NIOSH itself recommended regular rest periods and eye examinations for data entry and word processing people. It also recommended ergonomically sound chairs and workstations, with frequent radiation checks of the terminals, especially after repair.

How long and how frequent should breaks be? Five to ten minutes for every thirty spent working is optimal.

In any case, is it reasonable that we should chain the worker to the brain in the same tyrannical way as day laborers used to be chained to their bodies? Healthy, alive human beings need to find their own cycles of rest, activity (mental and physical), and refreshment. No one should be forced to input boring material for 7.5 hours a day, day after day, week after week, with no chance for monotony relief during the day. This is just not good for any living system.

- Screens should be tiltable.
- "Good" viewing distance is about twenty inches.
- Equipment using fans for cooling should vent the warm air away from the human operator.

- Adjustable copyholders placed behind the keyboard can save a lot of eyestrain.
- Amber-on-black and green-on-black screens are easier on the eyes than black-on-white or white-on-black.
- Flicker on the screen causes eyestrain. The brighter the light in the room, the more noticeable flicker will be.
- Glare can cause headaches. Reduce glare by careful placement of room lighting and by using the brightness and contrast adjustments on the screen.
- Keyboard placement is crucial. It may be useful to buy a special keyboard for your micro system if you spend long hours using it.
- The "standard" keyboard may not be the best. Look into alternative keyboard arrangements, such as the Dvorak design, to improve your productivity at the keyboard.

_____ 2.6 PERSONAL NETWORKS

Your personal network is part of your information territory. How you build it, expand it, relinquish it, share it, give it away, or use it will depend greatly on your values and goals. Personal networks penetrate and blend with your softspaces. Personal networks will consist of many other people and their softspaces, as well as the communications links between you and them.

All the networks you participate in, whether you designed them yourself or not, will be personal networks as far as you are concerned. The best way to picture the global network is with you at its center. Everyone else, of course, will also picture themselves as the center, and be equally correct in so modeling the world. In the months and years to come, the actual information transmitted through a particular network will be of secondary importance compared with the quality of the individuals who make the network hum. Greater value will be placed on how close one is to the originator of particular new information. To be "one of the first" or "the first" will carry greater rewards, prestige, and freedom.

Insofar as you can be creative, you can make original contributions to your networks, and making them that much more valued by your friends, associates, and clients.

A wonderful thing about working with information and networks is that leadership, innovation, and creativity cannot be monopolized, consolidated, or controlled by any one person, group, or nation. Today's have-nots can leapfrog into tomorrow's haves. Today's leader is tomorrow's follower. Failure—a necessary feature of growth in living systems—need not put you "out of the game," since the game is essentially ongoing and continual.

2.6.1   Personal Network Assessment: Checklist

The following suggestions can be used as a guide for assessing your personal networks.

- List the organizations you participate in: trade associations, professional organizations, clubs, special-interest groups. Be as exhaustive as you can. These relationships can be integrated into your network to the degree that they have access to or can be made available through telecommunications enhancement. They can function both as information sources and as destinations for your value-added information products or services.

- Diagram your personal friendship network: those individuals with whom you have daily or weekly contact and whom you consider to be part of your peer group. Figure 2.6.1 shows one way of graphically representing your friendship network.

- List the information resources commanded by each node of your network(s).

- List the information resources that you might be able to share with each of your current networks.

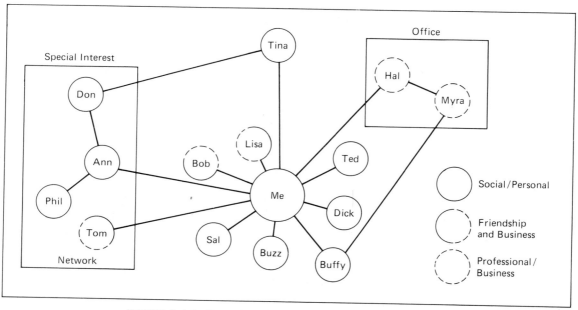

FIGURE 2.6.1. Sample Map of a Personal Network

- List the resources you need but don't yet have. Assign costs (in both dollars and time) to the acquisition of these resources.
- Do a "network market gap analysis" to find areas in which information networks are needed but not yet available, if at all.

2.6.2   Personal Netweaving

Personal netweaving involves several systems and systems levels.

- Your personal goals, values, and learning styles [2.0].
- Your hardware in your personal softspace(s) [4.0].
- Your software designed to run with your hardware [4.0].
- Communications links to others (phone, computer, cable, print, and so on) [4.0].
- Other people, their individual interests, softspaces, and related sub-systems [3.0].
- External data bases [3.0].
- Cross-connections, initiated by you or by others.
  The more cross-connections there are in your networks, the more resilient and supportive they will be for you. If they are initiated by you, you must be sure that you are not encroaching inadvertently on someone else's information territory when you make the connections. A good rule of thumb is to always ask if the other party *wants* to be connected to someone else, for whatever reason.

2.7   IMAGINING INFORMAGINATION

In the Information Age imagination is the key to personal and social transformation. Futuremaking is a function of communication, on personal and social levels. If true communication is possible only between perceived equals, as communications theory seems to suggest, how can we make a future with fewer bureaucratic strictures in our institutions?

It will do us little good to career toward the future with nary a glance at the dislocations and pains that misapplied and poorly understood technologies have already produced. This suffering may give us hints about avoidable scenarios to come, if only we pay enough attention. For example, if we have the potential to form highly innovative, creative networks, what good might they do if, simultaneously, we also create *scanning networks* whose job it is to scan the communications channels looking for "subversive" and "politically sensitive" transmissions? In recent times, this kind of scanning has been employed by national governments monitoring all transnational communications of individuals, corporations, and, of course, other governments. If you have sent a telegram or fax overseas, or made a transoceanic call, chances are that your transmission has been monitored and scanned for keywords by one or more government agencies.

Creative behavior is iconoclastic, unpredictable, and sometimes deemed "immoral" by self-appointed guardians of public sensitivities. Fantasy has a bad reputation among such people. If we are not to have monitors on our networks, looking for subversion, dirty words, and dangerous ideas or so-called "computer criminals," we must rescue fantasy and imagination quickly.

2.8   RESULTS: THE PRAGMATICS OF COMMUNICATION

The conceptual shift from an emphasis on energy to information in living systems is almost complete in contemporary models of human behavior. This shift is particularly and peculiarly valuable to individuals. As a netweaver, you can assist in the transfer of the new, Third Wave frames of reference in your relationship with yourself, your associates, and your work. *Information about an effect,* if fed back to the person or system performing the action(s) that produced the effect, can ensure both stability and adaptation to vast changes on personal and social levels. In communications terms, the netweaver will always be dealing with pragmatics: the results of a given communication or set of messages.

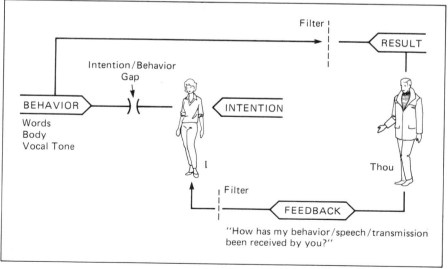

FIGURE 2.8. Dyadic Communications

Throughout your work as a netweaver, you'll be sending and receiving messages, setting goals and communicating goals, making decisions and communicating decisions. The most important messages will be, first, those you send to yourself. Careful self-observation and careful listening to the feedback of others will give you clues about the effectiveness of your communications. Monitoring how you feel, behave, and learn will help you streamline your own processes, and make communicating with others that much easier and more effective.

Figure 2.8 is a systems model of communication between two persons (a dyad). The model can be extended to include more than one person, institution, and so on.

If you send messages that result in outcomes you do not want and did not expect, then you must change your communications until you get the desired results. Whether you are acquiring new networking skills, starting a network-based business, selecting and installing networking tools, or just improving systems you've already designed, being fully in communication with yourself will smooth the path to your desired goal by helping in your communications with others. It will also help you avoid stress-related illnesses and other sorts of unhappiness.

3

Social Systems

Many social scientists have withdrawn into the mists on the mountain tops and can occasionally be seen mysteriously rotating factors and making multivariate analyses. Many instruments of science are there, and the incantations in jargon, but not the results. It is becoming quite embarrassing.

—Charles Hampden-Turner

The real world of the netweaver is no misty mountaintop. All our technology swims in a social sea. Sociology can be as practical as ship design or engineering. The engineer who has to get involved with "management" will tell you that in an instant. All our "schools of management" and books and magazines purporting to teach "management" are fonts of applied sociology. Even dismal economics is a branch of social science.

As soon as you step out that front door and into day care, nursery school, or kindergarten, you move through social systems of greater or lesser dimension. You are subject to school boards and public health officials, police officers, crosswalk guards, and teachers. Let's not forget the teachers!

Our social milieux, like our genes, are "given" to us willy nilly. For the first ten, fifteen, thirty, or one hundred years, we can do little to influence the social systems that influence us. Noticing that *some* people do come into the world seemingly well equipped to radically alter their social environments should give us hope. Gaining mastery over the tools of netweaving means that at some point we must learn to work with, within, and upon our given social systems and those we are trying to renew or invent anew.

_____ 3.1   PREHISTORY

Ancient humans knew and classified what they could see, hear, and touch. They differentiated their environment in ways as complex as modern scientific classification schemes. Out of sticks and stones the first tools were created. The stick was used to make marks in the sand. Rocks were everywhere, everywhere becoming sand. The shamans of the tribes discovered rules that helped differentiate rock from rock. "Here is ordinary rock," they said, "and here is flint rock."

Flint rock was more useful than ordinary rock. When attached to a stick, flint rock–stick system became a spear or a knife. Flint tools evolved along with the rules for making better and better flint tools.

With further rules for differentiation, metal was claimed from the rock of the earth. Metal was used for a long time before electricity, in turn, was differentiated out of it. By that time, the rules had become that much more valuable to humanity, and rules were discovered for discovering and creating more rules. These were metarules.

Soon, voices were being added to electricity and then differentiated out and returned to sound again. When voices are attached to electricity, they can travel at the speed of light. While voices were traveling on light waves, and while humanity as a whole knew a whole lot more about things other than what they could see and touch, they still behaved, by and large, as though the only "real" things were those that can be seen with the eyes and felt with the skin and heard with the ears.

3.1.1   Mass Memory

Without the simultaneous development of cheap, large-scale mass storage for digital information, the microprocessor would have been a much more limited tool than it has turned out to be. We could not begin to talk about the simulation of the human brain without also being able to simulate the phenomenon of memory. Magnetic storage of digital data is getting cheaper and more plentiful, even as more and more circuitry is shoehorned onto silicon wafers. The floppy disk drive soon will give way to laser videodisk and solid-state bubble memories. Beyond that lies the holographic storage of data, promising mega-mega memory caches for everybody. The videodisk wedded to microprocessors "shall make the crystals speak."

Human culture is made possible by human memory: the passing on of rules (information) for doing utilitarian things with and within the physical environment. The oral tradition depended on poetry as a kind of mnemonic aid for the retention of cultural information.

There are some scientists who claim that human biological evolution ended some aeons ago and that, now, evolution is primarily cultural in nature. Certainly the history of technology from the Stone Age to the present seems to support this idea. As a prime artifact of contemporary culture, the computer may be the next logical step in physical evolution. Certainly it depends on a host of techniques and tools of human culture for its current existence. Those who make such arguments can be found in the fields of artificial intelligence (AI) research, biology, and anthropology.

3.1.2   Memes

When they look for the cultural equivalent of genes, these scientists propose the existence of memes. A meme may be thought of as a unit of cultural information. Perhaps a meme is as small as a single concept or idea. On the other hand, it may be as large and as complex as a poem or computer program. In any case, memes are stored—remembered—in our books, our plays, our poetry, and our mass digital storage data banks.

Memes, like viruses, can be pathological. What seems to matter most is the "fit" and/or appropriateness of the relationship between a meme and its host system, whether individual, group, or society. Thus, for example, while it may at one time have been culturally and biologically beneficial to have a division of labor along gender lines, the memes supporting this pre–Industrial Revolution arrangement are no longer functional. They are pathological when considered in the light of our need for creative participation by as many minds (masculine *and* feminine) as possible in global problem solving and design. Contemporary liberation movements are devoted to wiping out dysfunctional memes just as health workers wipe out disease-bearing viruses. The challenge is to immunize ourselves against virus memes without coercion, brainwashing, or Big System paraphernalia.

Memes that are incorporated in engineered artifacts—technological memes, if you will—are especially interesting as a class. These memes seem to be most active, reaching out and grabbing one another, forming exquisite structures and novel configurations, making new memes from old. Microelectronic chips take larger and larger sets of ideas and package them in integrated structures. Netweavers take basic techno-memes—modems, protocols, programs—and incorporate them into unique information processing softspaces.

3.2   COMPUTOPIA

The Japanese have been examining the concept of "computopia" and its implications for some time now. They have leapfrogged most Western nations in understanding the systems dynamics between technology and Japanese society, including economics. The Japanese have instituted a nationally based effort designed to study and make recommendations concerning the Information Age.

Yoneji Masuda, a leading Japanese computopian, outlined some of the characteristic differences between information as a commodity and ordinary goods and services.

For starters, information is unconsumable. Ordinary goods disappear with use, but information remains essentially the same. As one wag put it, "Selling information is like prostitution: you have the product, you sell the product, and you still have the product." When I transmit information to you, you may have more (provided it was news to you), but I certainly don't have any less. Even so, I may give up a substantial advantage by giving you certain kinds of information that I hold.

Any given item of information must travel as an indivisible set. While physical goods can be divided and used, information, to be useful, must retain its integrity. It will not do you much good to have only half of a computer program, or only the top half of a book and not the bottom half.

Finally, information and its effects accumulate over time. This is where the creation of essential new wealth comes in. The quality of information can increase with use, rather than deteriorate. In this "anti-entropic" characteristic, information is most *unlike* other commodities.

3.2.1   Information Economics

How often have you heard someone say, "Our economy is a mess"? Economics is a perennial political theme and nemesis of politicians. Everyone who is not the president and/or not in a "healthy" sector agrees that "the economy" is sick. The diagnosis of the disease varies depending on the political ideology of the doctor.

Those who deal in the information-related industries seem, for the time being at least, to have built up some immunity to the spasms that clutch the rest of the economy. Even so, the 1983 corporate dropouts from the game of micro hardware manufacturing, such as Osborne Computers, Inc., and Texas Instruments, and the problems encountered by such "name" manufacturers as Atari and Sinclair, would seem to indicate that the microcomputer segment of the information industry is not immune to seizures and spasms.

How much, we might ask, are the problems in the *rest* of the economy being exacerbated by our transformation from an industrial to an information culture?

Any attempt to model our local (U.S.) economy is necessarily going to be complex. When using economic models, it is a mistake to assume that we can generalize from a small system, such as a small business or a family, to the macrosystem we call *the economy*. Econometric models are the most developed of any of our social systems models. Even so, they aren't very good. Models, no matter how sophisticated, can never incorporate, in their workings, *their own results.* In the real world, we use the results of models to modify our decisions, thus rendering the model obsolete.

Perhaps what is needed is a *system of models,* some of which incorporate real decisions of real people in their workings.

3.2.2   Information Technology

The underlying technology of information can be contrasted with industrial (Second Wave) technology in several ways. The idea of "economy of scale" cannot be profitably applied to information beyond a minor level. Basic physical laws (propagation velocities, physical size, cooling requirements, "housekeeping" needs, etc.) tend to limit the practical size of individual machines.

However, by linking machines into networks, a kind of "distributed machine" of very large capacity can be built. The telephone grid is the beginning of just such a system. As industrial tools such as power shovels and punch presses increase in capacity, they also grow in size and cost. As information machines increase in capacity, they are shrinking in cost and size. Therefore, information technology inherently moves us away from ideas of economy of scale, bigger is better. Centralization of information machines or information itself accrues no major benefits.

Thus, the idea of an "information utility" modeled on Industrial Age utilities is an anachronism. The idea exhibits the heavy, centralist, structural notions of industrial technology.

3.3   SELF

What holds your attention has your energy to the degree of the intensity and length of your attention. If you give your attention to a newspaper, the publisher of the newspaper has your energy *to some degree.*

Each of us depends a great deal on communications and feedback—the attention—of others in the creation/design of our sense of self. In turn, we express who we are using our entire neurophysical system to send messages about ourselves to others. Before 1970, human communication theory focused almost entirely on verbal forms. The use of language was the major study partially because it was thought that language was what set the human species apart from "mere" animals.

A systems view of human communication began to emerge beginning in the early 1950s. Study then shifted from *who communicates* to the *media of communication.* What we have now, when we talk about human communication, is the overriding idea of a *system of communicating behaviors* employed by members of a culture. This system in its entirety conveys coded meaning. Language merges with gesture, facial muscle signals, touch, interpersonal spacing (territory), pheremones (odor), writing, reading, listening, and contextual filtering.

Your communicating behavior is a gestalt—an unbroken whole—extending to your construction and use of networks. Much of your communicative behavior as a netweaver will serve to instruct or alter the ongoing communications processes of others. By examining the larger contexts and nested context levels within which your communications take place, you can alter your processes to achieve the goal(s) you've set. One proviso is that you consciously and deliberately make sure that you have feedback from the level of the system you are trying to impact. This feedback itself is a gestalt consisting of communications to you and information about the actual behavior of the larger systems.

Your own processes—communications to self about self—must always be the foundation on which the rest of your communications are built.

The-brain-as-computer uses four subsystem computers operating together, in parallel. Way down deep in the middle of the skull is the "old brain." The reptile brain. The snake. The alligator. The dinosaur! This biocomputer controls basic functions: hunger and thirst, sex, pleasure and unpleasure (fear, rage, pain), and smell: the main sense of the primitive creature.

Over the old brain, the rind, or *cortex*, grew. The cortex has two sides, and functions appear to be somewhat divided between the left and right. There are multitudes of connections between the sides, of course. In humans, one side or the other is responsible for language and language processing. Usually, it's the left side, but most left-handers and some cultures use the right side.

Finally, the cerebellum—"little brain"—handles muscle coordination and is sometimes called the autopilot. It can take over from the cortex in accomplishing physical things like bike riding or automobile driving while you think of something else with your cortex.

Lower in the brain stem, as the spinal cord enters the head, are many of the preliminary filtering systems responsible for attention.

Over a lifetime, the human biocomputer produces 10^{11} bits of information, more or less. Current practical technology allows us to move information at the rate of 10^{15} bits per second. If a complete record were made of your lifetime's information output, it could be zipped into and out of Central Archives in about one one-hundredth of a second. Your primary "data output subsystem" is your voice. Your voice can now be treated just like any other data source: text, music, and images. It can be collected, processed, stored, transmitted, used, and reused just like any other data. When is your voice a "private channel" and when is it "public"? If you work for a company, will all your verbalizations during a given working day be treated by the MIS department as "a new and important company information resource that needs to be managed"?

A given utterance or communications output from a netweaver can have, as its source, any one of the brain's centers outlined above. It is important to stop and ask, on occasion, "Which part of my brain am I operating with at this moment?" Most systems of self-development are designed to balance the working of the entire brain, to integrate all the brain centers into symphonic rather than arbitrary staccato output.

3.3.1   The Culture of the Entrepreneur

Some mainstream economists estimate that the United States must "create" 20 million jobs by 1990 in order to be able to provide a job for all who want to work. No matter what the definition of "full employment" is, it is unlikely that "they" will be able to create that many jobs. What is needed is for people to be able to create jobs easily for themselves. More important, people must be able to *see themselves as capable* of creating their own employment. In the Information Age, this is not only easier to do, but is *the* essential challenge for society as a whole. The rate of unemployment in 1983 amounted to (conservative estimate) 9.5 percent of the "potential work force." "Eradicating joblessness" means reducing that percentage to 4 percent or less of the nation's work force. Since 1976, only 17.5 million new jobs have been created.

Jobs in heavy industry—the main source of employment in the industrial economy—are disappearing. New jobs require new skills, and universal lifelong education of humans is not as widespread as robotic retooling of factories.

The task of preparing people for new ways to work is one of the major needs of the Information Age. It appears that it will be an ongoing need. The same solutions will be helpful to those put out of work and those wishing to change jobs or careers. Business will spend $300 billion over the next ten years in its ongoing training programs. These training programs, in turn, will spawn their own new jobs and new approaches, especially as telecommunications and media technologies make possible the widespread distribution of otherwise specialized information.

Who will create the new jobs? In our capitalist culture, the role of the entrepreneur has emerged as a major factor in creating new jobs. The entrepreneur is a self-starter. Restless, energetic, passionately committed to a given new idea, the entrepreneur is a *champion* who may, in fact, sacrifice a great deal for a particular idea in order to see it materialize as a concretely accomplished goal. The entrepreneurs who have been identified and studied by social psychologists share many traits in common with people who have been identified as "creative."

There are some signs that an "entrepreneurial culture" is emerging centered around and supported by the new information technologies. The people participating in creating this culture are motivated more by mental than by sensory satisfactions, more by ideas than by "bottom lines." Their goal is not to own their own business but to see a goal realized, which requires that they start their own business. In the process, new jobs are created.

3.3.2   Systems Analysis

In systems analysis and systems design, a system is regarded as "an ordered set of methods, procedures, and resources designed to facilitate the achievement of an objective or objectives." If a system is already in place, system analysis will consist of a data-gathering phase, an analysis and recommendation phase, and an implementation phase. If there is currently no system in place, then a systems design can be worked out, beginning with a clear statement of the goal(s) and objectives of the system, with appropriate prioritizing and "weighting" of the required outputs of the system.

Even though it may appear to an analyzing observer that there is no system, this may not be the case. For example, many people may claim that they don't have a time management system for themselves. On closer examination of their values and actual day-to-day behaviors, however, the system they actually use will emerge. What's more, this system may work perfectly well in terms of the individuals' needs and actual performance. In fact, before attempting to introduce any changes in self-systems, netweavers would do well to simply *observe* and *record* their day-to-day behaviors for a period of time. For example, if you want to make changes in your personal economic system, begin with the discipline of recording what you are doing with money from day to day. This self-observation then becomes the self-feedback you use to make the changes you need to make. Likewise, in the management of time, begin with how you actually treat the time allotted to you in a twenty-four-hour cycle. Record your sleep cycle needs and your daily performance cycle. Before you can effectively manage time, you need to know what your daily rhythms (called *circadian* cycles) are now. Changes you make can then be brought in line with your needs, which makes it more likely that improvement in your time management abilities will happen.

In the formation of small groups and individual businesses (sole proprietorships or partnerships or small corporations), a system emerges and begins to evolve as soon as the people involved begin to communicate with one another. If the initiator of the system is a single individual, the system will bear the biases of that individual for a long time after its initiation.

A netweaver who is called in as an outside consultant to an ongoing organization would do well to "do nothing" but observe and record before rushing in to change or recommend or advise.

Reaching a goal involves problem definition and problem solving. If the goals are not well defined, the initial survey or data-gathering phase can help determine that. This is sometimes also called a *feasibility study.*

This phase may include experimentation with a small-scale pilot system in order to find out more about the systems behaviors that result. For example, if you are setting up a communications network for a business, you may want to begin with a small, easily used system, including a representative set of people within the company and technically simple hardware to help determine actual usages and performance of the overall design.

3.3.3   Systems Design

After the survey or data-gathering phase has been completed, you should have a better idea of what the problems to be solved are. Moreover, you'll have a better idea whether they *can* or *should* be solved. (Don't forget that thorough analysis will include individual *needs* and value structures at individual and group levels.) The solutions, insofar as they involve actual requirements for hardware, software, and organizational connections, will satisfy the requirement list generated in the analysis phase. These requirements are then the basis for your design.

What functions are needed in the system in order to satisfy the requirements? Answering that question, in detail, leads to a detailed set of *specifications* for the new system(s).

The design checklist (in order) goes as follows:

- State the problem(s). (Create a network for XYZ purpose.)
- State the requirements. (Must be representative of the community served. Must be easily accessible. Must begin with XYZ information.)
- State the constraints. (Must be within $dollar budget. Must begin with those currently interested. Must begin with currently available information. Must meet specified needs.)
- Do a survey. (Is there a need for an XYZ network? Can one be built? Who is currently interested or involved? What is the community setting? Are the potential participants able to use the available technology? What are they doing now to meet their needs? What do individuals and the various subgroups within the organization want the network to do?)
- Do an analysis. (Prioritize both needs and constraints. Identify potential hardware, software, and people resources. Make a "wish list" identifying quality characteristics of the new network in detail.)
- Design a system. (Identify and specify hardware, software, participants, membership or participation requirements, entry procedures, if any.)
- Implementation. (Evaluate vendors. Get third-party referrals. Buy the components needed. Set up and run the hardware/software. Get the people involved. Do necessary training. Monitor performance.)
- Feedback. (Get user feedback.)

Note that at the initial stages of the design process, a single individual can do what is required. If the network is highly personal and involves your close family, friends, associates, or learning colleagues, there is a greater likelihood that you can implement most of the phases of system design and implementation on your own. However, if you want *participation* in the network that results, it is a good idea to bring as many of the potential participants into the process as early as possible. In fact, if you can get your potential participants to state the problem themselves, it is more likely that they will become involved in its eventual solution. Likewise, if you attempt to state the problem on your own, you may get resistance from people who have not identified and will not acknowledge the problem *you* see or state. It is even less likely that they will participate in making the network emerge. Without them, there *is no network.*

3.3.4   Privacy

At the individual level, we can talk about four states of privacy:

Solitude: Physical separation from the group.

Intimacy: Participation in a dyad, triad, or larger group which itself achieves corporate solitude.

Anonymity: Freedom from being identified and/or placed under surveillance while in a public environment.

Reserve: the creation, through our individual communications systems, of a psychological barrier or boundary that serves to protect us from unwanted intrusions.

Privacy is needed before personal autonomy can be attained. Self-evaluation and emotional release also require individual privacy. Privacy requires that each of us be able to *limit and protect* our communications with ourselves, significant individual others, and groups. This means, by extension, that our channels of communication at all levels be free from intrusion and/or surveillance, whether by people or by machines.

The functions and requirements served by privacy must be an integral part of any network design. This may include data protection, encryption, and system access regulation.

The "right to privacy" is essentially and fundamentally an individual right. It can only partially and incompletely be extended to groups, government bodies, and corporations. This is because the requirements of accountability and the public "right to know" reduce, in absolute terms, the demands of privacy by multiperson institutions.

Along with a universal individual need for privacy, there is also seemingly a universal tendency on the part of individuals and groups to invade the privacy of others and to engage in Big Brother tactics to enforce norms, whatever those norms may be. Curiosity, gossip, the quest for "explanations," and the thrill of vicarious experience become, at higher systems levels, dossiers in data banks, credit reports, industrial espionage, and "law enforcement." Obtaining *any* information about people against their wishes constitutes an invasion of privacy. This includes "derived" and/or inferential information derived from legitimately collected data, such as psychological profiles created from an individual's buying patterns as reflected in economic transactions, say in a credit card data bank or in bank checking account records.

The protection of privacy on all levels requires, in the Information Age, that we employ both technological design solutions (embodied in hardware and software) *and* social/legal protections. Above all, we must become more sensitive to the *potential* uses as well as the stated, actual uses of collected data.

The ways in which privacy can be abused are legion. They include, but may not be limited to:

- "Ordinary" searches and seizures with or without a warrant
- Electronic monitoring: phone taps, remote microphones, recorders, etc.
- Coerced disclosure: information released under duress, threat of imprisonment or criminal charges, and governmental record keeping or registration (the census, the military draft)
- Spooks and spooking: secret agents, informers, disguise, ploys and trickery
- Observation of public events and the recording of transient information: photographing crowds at demonstrations and rallies, the piecing together of intelligence data to create "new" information
- Payment and/or an actual or promised reward or benefit: giving your name and address to a marketing company in return for the promise of winning a contest
- Disclosure of personal information to a third party by a legitimate agency without the consent of the subject of the information: bank records turned over to IRS agents

The Privacy Act of 1974 was passed in order to erect some barriers to abuses and invasions of individual privacy.

3.3.5   Libel and Slander

Of course you want to protect your own privacy. But what about your neighbors' privacy? For that matter, what responsibilities does a net-weaver incur along with new communications abilities? U.S. laws covering defamation of one's neighbors go back to common law inherited from England. Defamation includes *slander*, which is anything you say with your mouth channel, and *libel*, which is anything you say in print, on a broadcast medium, or on a micro telecommunications system.

Netweavers should arm themselves with a knowledge of what they can and cannot say in print or on videotext media. A libel is any false statement about a person that tends to bring him into public hatred, contempt, or ridicule, or to injure the person in his business or occupation. Those who sponsor microcomputer conferencing systems (system operators, network facilitators) can take steps to ensure that libel suits don't happen as a result of the use of their communications facilities.

- If a statement can be construed as injuring someone's reputation, it's potentially libelous. Get rid of it.
- Post policies on your system(s) stating what libel is and under what circumstances such statements will be deleted from your conference data base.
- If a statement is factual, but still potentially injurious, make sure that the facts are true. Truth is a defense against accusations of libel, but proving truth in front of a jury is difficult.
- Check and doublecheck the facts. Get second source confirmation, just the way a competent newspaper reporter does.
- Personal opinion based on fact allows greater leeway in one's utterances. Personal opinions clearly stated as such cannot, generally, be held to be libelous as long as the opinion is based on facts.
- Never, ever make use of names or pictures of people to promote anything, commercial or personal, without getting *explicit permission* to do so. Counterculture promoters are notorious for using high-power names in contexts that would appall the bearers of said names.
- If you do get into trouble, and someone accuses you or your network of libel, never publish a retraction without good legal advice. Don't say anything at all, in fact, until you can determine, with the help of an attorney, where you and your system stand and how vulnerable you are to the accusation.

3.3.6   Human Rights

That there are human rights is a contemporary form of the doctrine of natural rights, first clearly formulated by the philosopher John Locke and later expressed in terms of "the rights of man." Natural or human rights are rights people have just because they are human and they are on the planet, not because of government fiat, law, or convention.

Such rights have frequently been invoked in criticisms of laws and social arrangements. In 1948, the General Assembly of the United Nations adopted a Universal Declaration of Human Rights. This declaration formulated in detail a number of rights: political, cultural, and economic. This became a standard by which nations are measured. Of course, this was not a legally binding instrument, but it was followed by a number of international covenants and conventions, including the European Convention for the Protection of Freedoms, which have influenced national legislation and provided some machinery for international enforcement.

The defense of human rights on a planetary scale must include defense of civil liberty and human liberty. Guaranteed rights would include the right to good nutrition and health, the right to know the truth about any given subject(s) or to seek the truth if it is unknown, sexual intimacy, leisure and relaxation, beauty and travel, the right to be unique. Human beings are not "civilized" in any sense of that word without the right to be whole. Above all, the right of each individual to control her own body and CNS, including the mature right to alter one's consciousness according to one's own needs and desires, must be paramount. All these rights can be further enhanced and protected by a right to be protected from ignorance, one's own or someone else's.

Information systems and information exchange networks can help bring these rights to humankind. Neurological autonomy is the cornerstone of a Planetary Bill of Rights.

3.3.7   Transforming the Bill of Rights

Amendment I

Congress shall make no law respecting an establishment of religion, or prohibiting the free exercise thereof; or abridging freedom of speech, or of the press; or the right of the people peaceably to assemble, and to petition the government for a redress of grievances.

As more and more human transactions are mediated through micros and terminals connected to the phone lines, freedom of speech and freedom of the press begin to merge. The citizen's right to secure channels of communication, free from prior restraint, censorship, eavesdropping, or machine monitoring should be built into the various layers of the electronic systems involved. The ability to organize politically, via computer conferencing and networks, would be safeguarded thereby, in conformance with Article I of the Bill of Rights. Here, the right to assemble peaceably is broadened to include people "assembling" in electronic softspaces with others on the network.

Amendment IV

The right of the people to be secure in their persons, houses, papers, and effects against unreasonable searches and seizures, shall not be violated, and no warrants shall issue, but upon probable cause, supported by oath or affirmation, and particularly describing the place to be searched, and the person or things to be seized.

Authorities should have to get a search warrant in order to enter private or public data bases for purposes of copying information that belongs to a citizen or is about that citizen. The specific kind of information sought, and the reason for the intrusion, should have to be spelled out under oath. Being "secure in one's effects" should include portions of memory maintained by someone else but being rented by an individual. It should also cover personal hardware, software, and peripheral support material, such as disks and tapes. "Effects," in other words, should be broadened to include information as well as physical artifacts.

Amendment V (partial)

No person . . . shall be compelled in any criminal case to be a witness against himself, nor be deprived of life, liberty, or property, without due process of law; nor shall private property be taken for public use without just compensation.

This should cover such things as banking records, personally generated encryption codes, records of transactions with data bases outside the home or office, and so on. The concept of "private property" should be expanded to include the contents of one's mainframe data files. In addition, it should be applied to prevent "fishing expeditions" by agencies such as the IRS and NSA, who have computers large enough to "raid" and correlate information from many different computer sources on private citizens. Such unauthorized information raids and correlating activities should be hemmed severely by both technical and legal fences.

Amendment VI (partial)

In all criminal prosecutions, the accused shall enjoy the right . . . to be informed of the nature and cause of the accusation; to be confronted with the witnesses against him. . . . .

Amendment VI has obvious implications in the credit data field. Where livelihoods and the ability to participate fully in a computerized society depend on accurate information, an individual should have the right to see *any* dossier or file created on or about him or her. Further, we should have the right to challenge the veracity cr accuracy of such information, and to get it changed in court if need be. The Freedom of Information Act and the Privacy Act are steps in the direction of such safeguards.

Amendment IX

The enumeration in the Constitution, of certain rights, shall not be construed to deny or disparage others retained by the people.

"Retained by the people" implies the use of microcomputer technology by private individuals, and the creation and use of high-security encryption techniques by anyone who wishes. One should not require a license from the state for either.

Netweavers can help in the development and protection of the Bill of Rights in the Information Age by adopting the following ethical guidelines and practices:

1. Netweavers will encourage and support the widest possible public participation in decisionmaking concerning telecommunications development, uses, and policies.

2. Netweavers will encourage and support maximum public use and benefit of new and existing channels of information, data transmission, data bases, satellite transmission technologies, and hybrid micro/video/telephonic technologies.

3. Netweavers will encourage and support the rights of private individuals and private groups to research and develop new data encryption technologies, including but not limited to both theoretical and applied cryptographic research, applied encryption techniques in hardware/firmware and software form, and the development of public-key cryptosystems.

4. Netweavers will uphold and maintain the rights, as supported by the U.S. Constitution and the United Nations' Universal Declaration of Human Rights, of private citizens and groups to secure channels of communication and information, free from any and all forms of monitoring, whether by governments, other private groups, or individuals.

5. Netweavers will encourage and support the development and spread of electronic communications systems that actually incorporate—in hardware and software—the privacy and security features implied by the Bill of Rights.

6. Netweavers will be aware of and communicate to others an awareness of actions and pending actions of government agencies, legislative bodies, private groups and individuals which might have an effect on the overall freedom and security of information channels, via present or future communications technologies, such that human rights guarantees are further secured in the emerging global information environment.

7. Netweavers will collect, process, and share information with one another relating to all the above, preferably on an international basis and

with a view toward the development of a Planetary Electronic Bill of Rights.

_____ 3.4   GROUPS

Skim through current group handbooks, conference leaders' tool kits, and the like, and you find what sounds very much like a call to arms by the mediocre against their enemies.

—William H. Whyte, Jr., *The Organization Man*

When we began our research, we originally thought of the group as a means of making the creative process visible so we could examine it; the meeting itself, the way people worked together, grew in importance in our concern until it more than equaled the procedures for developing ideas. Because of what we heard and saw happening, we have had to question many basic assumptions about meetings and, in particular, meeting leadership. Many observations surprised us; their sum strongly suggests that the traditional problem-solving meeting is a blunt instrument, not an incisive one.

—George M. Prince, *The Practice of Creativity*

A Quality Circle is a group of workers, doing the same or similar work in the same department or work area, who regularly and voluntarily meet (usually once a week on company time) to discuss problems associated with their jobs. The group's objective is to identify the *real cause* of a problem and to discover a way to eliminate that cause.

—Bureau of Business Practice, *Quality Circles: A Dynamic Approach to Productivity Improvement*

Collective creativity involves multiple input, continuous feedback and group interaction in an ongoing manner all the way through the [design] process with no one person "doing the design."

—Halprin and Burns, *Taking Part*

3.4.1   Meetings and Conferences

The first computerized conferencing system was created in 1970. It was set up by a government agency, as one might expect. The Office of Emergency Preparedness (OEP) system evolved to become EMISARI (Emergency Management Information System and Reference Index).

EMISARI was supposed to change over time in order to meet the changing needs of its users. It included, among other things, a bulletin board feature, news files, and provisions for feedback from users to administrators. Some of the early history of this system is contained in *The Network Nation* by Starr Roxanne Hiltz and Murray Turoff. Hiltz and Turoff describe the early implementation of the Department of Defense's Advanced Research Projects Agency (ARPA) and ARPANET, a since-dismantled data communications network that linked university and government computers across the country. The net is essentially a military/academic conglomerate linking scientists who work directly or indirectly on military projects.

As microcomputers came into widespread use in the late 1970s, ARPANET was one of the first networks traveled by hobbyists, phone phreaques, and hackers intrigued by the challenges of accessing strange computers at odd hours of the night. The host node at the Massachusetts Institute of Technology was an especially popular entry point into ARPANET for these midnight marauders. This popularity was supposedly due to M.I.T.'s tolerance of "tourists" on its system. Owing, in part, to this "invasion" of ARPANET by microcomputing info-tourists, the Department of Defense "erased" ARPANET in 1983.

Information on ARPANET was unclassified, so there was no question of outsiders accessing supposed government secrets on it. ARPA had devised alternate networks for classified research and implemented MILNET for military use. As an entity, ARPA has probably done more research in network communications than any other agency in the world, public or private. How much of this research will benefit taxpaying citizens in the long run remains to be seen. One report put ARPA's spending on software alone at $3 billion per year. The result is that it has become ridiculously easy for ARPA to connect different computers, to exchange data almost instantly, and to create and dismantle different networks practically at will, regardless of the lack of industrywide—much less international—standards. All this capability has emerged without much public scrutiny, oversight, or awareness.

All this just to cut down on the expense of shipping people to meeting places. From there, it follows that substantial savings could be had from cutting down on commuting from suburbs to city. A netweaver can find plenty of opportunity to cut costs further by looking at business trips within cities. Studies in the United Kingdom point out that between 40 percent and 60 percent of personal travel within cities could be cut by using audio conferencing, video conferencing, and computer conferencing.

3.4.2   Decisionmaking

The effectiveness and health of organizational life depend in large part upon skillful group work.

The group is effective in bringing out points of view that might otherwise go unexpressed.

Together, the members of a group can usually see more possible lines of action than if they were polled individually.

Genius cannot function in a vacuum. Interaction with others in a particular field is not only stimulating but indispensable.

However:

Association with a group may not be vital to the task at hand. It may also be repressive.

To view groups as wellsprings of creativity beyond the capabilities of individuals within it may lead to "false collectivization." The arts of discovery cannot be tamed by any group or policy. Order, agreement, and the statement of objective goals may be necessary to the *execution* of an idea, but not to its *creation* in the first place.

Really new ideas are an affront to the systems—groups, organizations—within which they are introduced. They wouldn't be new otherwise.

3.4.3   Consensus Building

Before setting up a given network, it may be useful to hold one or a series of face-to-face, old-fashioned meetings with those who will be using and constituting the network. In the organizing phase, particularly with local community networks, it is important to build a consensus concerning the purposes, goals, and objectives of the network-in-formation. There are some operating principles that can assist the netweaver in getting through meetings with outcomes that work. Some of these principles are transferable to later work, which may be done completely online.

Most people hate meetings, or at least profess a disdain for them. Some meetings, indeed, like some parties or dinner dates, can be trying ordeals. But they don't have to be so. The quality of a meeting held among members of an already ongoing organization can often serve the netweaver as a litmus, signaling problems of the whole organization: imbalance, misinformation, or misalignment.

Meeting results can be improved if we just try to remember and come into tune with the dynamics of the small group in a "meeting mode."

Open the meeting with a time of silence of a minute or two. Ask individuals to focus quietly on *their* purpose(s) for coming together, to listen for the inner voice for direction and guidance.

Allow time for self-introduction, after the silent period, for people to begin to get to know each other. Insights generated during the time of silence can be shared at this second stage as well. This helps unify the group and its purposes. It affirms a commitment to honor the needs and intuitions and expertise of individuals, the source of all true creativity.

After the opening introductory phase, an agenda should be made final for the group. The initiator of the group, perhaps you, might have drawn up a preliminary agenda and distributed it to the participants beforehand, but be prepared to drop some items or add some items based on the input from others during the opening phases of the meeting. A useful agenda should specify:

- The content of the meeting,
- The process(es) to be used,
- The priorities of the meeting (which items should be covered first),
- The amount of time allotted for each item.

Some quasi-formal roles within the group have been found helpful in some cases, depending on the tasks of the group. In a group that has been going for some time, these roles and their attendant responsibilities can be shared around. A *synthesizer* or *focalizer* looks for points of agreement and restates them in the context of the "big picture" or "whole system." This person helps the group stay on track, reach decisions, and plan for implementation, and generally monitors communication. She suggests processes to help the group move forward to its tasks and goals.

A *fairwitness* or *mediator* pays attention to both individual and group needs. He makes sure that everyone is heard and reminds the group of the value of each person's contribution. If there are conflicts, between the group and a single individual or between individuals, the fairwitness helps mediate and resolve them.

A *transcriber* or *secretary* or *recorder* captures all the key ideas. Of course, if the meeting is an electronic one, conducted on-line, this role will be played by the host computer. The computer keeps the record of the meeting, which can then be edited or transcribed on agreement. In face-to-face meetings, this function should be performed in public. That is, rather than keeping notes privately, the recorder can keep notes on large pieces of newsprint taped to the wall or on a chalkboard. This kind of "group memory" is really inexpensive and worthwhile.

The *participants* are all co-processors in the group, combining their biocomputing power with that of the others. Participants help keep the other roles reminded of their roles.

During the meeting, whether for an hour, a day, a weekend, or on-line for the next six months, results can be improved even further. Getting three basic agreements right off will cover a lot of ground very quickly:

1. The purpose of the meeting,
2. Procedures for discussion,
3. Decisionmaking methods.

Each meeting, at the beginning, should get agreement on its purpose. If the meeting is an extended one, the purpose should be brought up from time to time and discussed. The focalizer or facilitator should be the one to do so if no one else does. *Purpose* here is synonymous with *expected outcome* of the meeting. A common understanding of what work is to be done will go a long way in getting that work accomplished. This is the "what" of the meeting.

Not only the "what" but the "how" of the discussion—the actual processes to be used—should be agreed on by the group. Will we work as a whole, or break up into committees and task forces and subgroups? Will we meditate and write, or brainstorm all possible ideas? The methods and varieties of techniques for creative problem solving in groups are highly varied. Some will work better than others for given tasks.

Finally, how will we arrive at decisions? Will we vote? Will we delegate decisionmaking powers to subgroups or individuals? Will we keep generating solutions until a consensus is reached? Consensus decisionmaking tends to honor each person's needs and input, but also takes longer than a straight-out vote.

If the meeting gets bogged down, or if some simple item seems to be getting too complicated in the group process, there are some simple techniques you can try.

- Split temporarily into smaller groups to discuss the issue. Reconvene after a set time and communicate the results from the smaller groups to the entire group.
- Be quiet. Take a minute or two of silence to help refocus the group.
- Ask questions. Is this relevant? Is everyone being heard? Is this the right time? Is our focus clear?
- Brainstorm. During brainstorming, only clarifying questions and new ideas are allowed: no criticisms, no "yes, buts" until all the ideas have been generated.

A good meeting results from a process that, at its best, *can* be better in its creative results than any generated by a single individual.

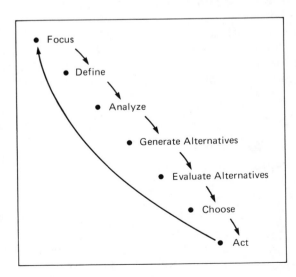

Often, meetings falter at the "choice and action" phase, partially because it is here that people have to actually commit themselves to the purpose of the group. Choice and action are often great problems for individuals who are brought into the group. Choices should result in an *action plan* complete with individual names, dates, milestones, and outcomes specified. This plan then becomes the group contract, ensuring that the meeting will result in movement down the road, and not just another waste of valuable time. After the meeting, or just before it adjourns, evaluate the meeting itself in the group. Was anything accomplished? Are there "agenda hangovers" to be taken up at the next meeting? Be silent again. Listen to the earth. Continue.

3.4.4   Information Exchanging

Meeting with people who have common interests can be stimulating. It can also suggest new approaches to one's own work. But stimulation may not lead to creation, and it is wise not to confuse the two. Meetings of minds can be profitable if one's expectations are not so high as to demand a creative result from such meetings.

Usually, it is not a good idea to call a "networking meeting for the purpose of sharing information." For one thing, everyone who comes to the meeting will probably have a slightly different idea of what "networking" really means. (Unless, of course, they are themselves netweavers and have read this sourcebook.) For another, "sharing information" is also too broad to have any sort of action outcome from the meeting. Focus is required: What kind of network? What kind of information? Who would be most interested? What outcomes do the organizers expect from the meeting?

People "exchange information" all the time. In netweaving, we want to accomplish more than good vibes and the exchange of business cards and a promise of lunch on Tuesday. In a start-up situation, we want to get energy behind the grounding of the network. What do I mean by "grounding"? More or less permanent connections between people, using netweaving skills to create an ongoing, active, and continuous exchange of information of *specific* kinds, for *specific* reasons, to serve *specific* ends and a *specific* community.

3.5   THE BIZ BIZ

If the futurists, analysts, marketers, industry witnesses, observers, and pundits are right, we are collectively on the verge of the creation of thousands of new information-related businesses that never existed before. That means support services for the creation and maintenance of start-up businesses are going to be more important than ever in the months and years immediately ahead. The "business business" is that group of products and services that make starting and running a business easier.

The need for telecommunications- and computer-oriented people to guide the start-up business into the Information Age is more vital than ever. A checklist of items to be aware of in the beginning follows.

- Ideas for new businesses are best developed from your own previous experience and skills.

- Keep your initial cash investment small. (Your idea of "small" will vary, certainly, with your income and assets.) In information-related businesses, it is generally easier than it might look to begin on the proverbial shoestring. And if you let lack of capital deter you from starting, you might never begin. On the other side of this coin is the overoptimistic: without a preliminary budget and some idea of the actual costs involved in your new business, you may be steering— more properly *not* steering—your baby enterprise right off a cliff. Bernard Baruch's advice is probably best here: never invest enough to lose sleep over.

- Information-based products are generally associated with low manufacturing costs. Stick with these rather than, say, hardware or peripheral projects.

- Subcontract where possible. This widens your network, does for your business what delegation does for the executive of a large company, and can result in a better product overall. Provided, certainly, that you pick your subcontractors carefully.

- Try to carve out a proprietary niche, where possible. [See *Positioning,* section 3.5 of Appendix III.] Use existing copyright and trademark laws to establish your business identity and image. Remember, even the *National Geographic* has been ripped off by counterfeiters and pirates. Don't make their job easier. Try to outthink them before you are successful and thereby a more tempting target for look-alikes and clones.

- Establish relationships in your market. Repeat sales and updates are the mainstays of information businesses. Remember that what you may lack in size must be made up in personality.

Many people are already actively working in their own "electronic cottage industries" at home or in small offices in low-overhead districts. These people are using the emerging computer and telecommunications grid to support their work.

The "at-home" business has generally been associated with the small-time, one-of-a-kind efforts of grandmas making pin money with crocheting, brownie baking, or crafts. Those who deal in business prognostications now estimate that, by the 1990s, more than 15 million people will be working out of their homes or close to them. About a third of these will be "telecommuters." Telecommuters will work for large companies, for the most part, on special projects that will be highly self-directed. Ten million or more will be working at their own self-guided businesses.

Even today, in 1984, more men are working at home because of the economics of renting business space in expensive high rises. Some also wish to be more active parents. And let's face it: no one would believe a book on the "Joys of Commuting."

While others active in the micro revolution nad seen the graffitti on the wall long before him, Alvin Toffler spelled it out in *The Third Wave*: the cost of moving information on the networks—telephones, satellites, cable, dish, micro synthesis—soon will be lower than the cost of transporting a worker from point A in space to point B.

Some potentially grave social problems are associated with this, though. The ability of large companies to exploit the individual worker, paying less-than-scale wages for piecework such as data entry, for example, concerns many observers. The lack of benefits and "wage security" are also some concern.

Working at home or for oneself is not for everyone, certainly. The task or purpose of the business must be appropriate. (No punch presses in the den, please.) The individual must be able to work well for long periods of time without direct supervision. The task must have, as its end result, a saleable product or service.

Marketing may be an insoluble problem for newly liberated Third Wavers. The idea of *selling* oneself or one's services to others is repugnant to many, particularly people who are overly "modest" about their talents, skills, and abilities generally. (Self-esteem, anyone?) Accounting and legal questions can be even greater hurdles for the independent home entrepreneur or "kitchen table business."

There are many resources available for the electronic cottage information business. Networks designed to help would-be self-starters are springing up all over the country, locally and nationally. If your area doesn't have enough of such resources, consider making your first net-weaving task that of creating them.

3.5.1 On-Line History

Much of the on-line information industry of the 1980s evolved from the remote data processing services (RDP) that began in the 1950s. These services, also called time-sharing services, were developed as an alternative to buying or leasing a mainframe computer.

The primary services performed by time-sharing companies involved payroll preparation, billing, and other accounting services. By leasing or buying a relatively low-cost terminal, either from the time-sharing company or from one of the terminal manufacturers, a smaller business could have the advantages and increased productivity associated with computerization without the direct costs of maintaining the computer.

Time-sharing is one kind of remote computer use through the phone lines. The other kind of use, becoming much more widespread today, is remote data base use. This kind of information retrieval parallels the development of time-sharing systems. The telex network was developed in the 1920s and was used as a time-sharing network as early as 1966 on a dial-up rather than dedicated line basis. In 1970 the National Library of Medicine entered into an experiment called AIM-TWX. TWX (pronounced *twix*), like telex, is a service of Western Union. It is sometimes called Telex II. AIM stands for Abbreviated Index Medicus, and the experiment provided information to the medical community.

Use of telex and TWX for computer access has been reduced to almost nothing these days, since the telex network transmits six characters a second, and the TWX network is only marginally better at ten characters a second, and, besides, is only available in the United States. However, in some parts of the planet, telephone service is highly unreliable, or doesn't exist at all. In some of these locations, telex is used because it is the only viable way to access remote data bases and other kinds of on-line service.

ITT and RCA GlobeCom have created gateways [4.21] through telex to TYMNET and Telenet, the newer data networks. These gateways convert the Baudot code [4.7.2.3] used in the telex networks into the ASCII [4.7.2.1] code used by the more modern value-added networks. As you might imagine, it's s-l-o-w . . . but it does the job.

3.5.2   Business Intelligence Increase

To deal successfully with information in conducting business, there are certain minimum skills that are not only useful, but perhaps necessary if the task of the business is to be accomplished. The designer or organizer of a new business should keep the following areas of concern in mind.

- Administration, office management
- Audiovisual communications
- Budgeting, planning, cost/benefit analysis
- Data communications/telecommunications
- Information processing
  Word processing
  Data bases (in-house and external)
  Electronic spreadsheets

- Ergonomics, human factors
- Micrographics technologies
- Organization development
- Primary organization expertise (e.g., skills in the main business purpose or task)
- Purchasing, procurement
- Records management
- Financial management
- Marketing/sales/public relations/advertising
- Software development and acquisition

This list is an expanded version of the "classic" critical business functions of marketing, innovation, production, sales, finance, and general management. (Two kinds: the kind you need to operate—metabusiness—and the kind that concerns the product or service of your business—data bases.)

An information planning task force in a business should have the following ring structure:

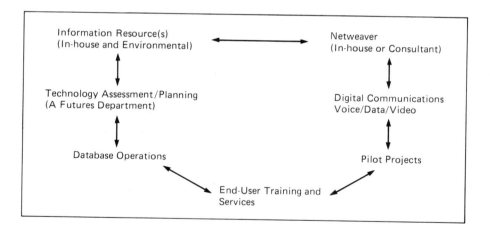

When doing a systems analysis of an ongoing business, the idea is to increase the internal intelligence of the organization through in-house data gathering and feedback. The following checklist covers some of the important areas that an in-house systems analysis should look at.

- Borrowing costs, if any. Cash flow management.
- Collection on receivables, account aging, if any.
- Invoicing processes and throughput. Are there delays in the system?

- Shipping process and throughput.
- Invoicing and shipping errors.
- Projected business volume growth? Too much too fast means as many problems as not enough too slow.
- Staffing adequate? Future needs for personnel?
- Productivity areas: What are they and which could be increased?
- Could clerical work be reduced or cut entirely in some areas?
- Information security: How much is there? How much is needed? Why?
- Audit trails: Especially in electronic systems already in place, are they adequate?
- Visibility: Does the business know itself, its community, its vendors and its customers?

Communications and information management can have an impact on each of these areas, for better or worse. It is more likely to be a positive impact if the initial analysis takes a little more time and is just that much more thorough as a result.

Figure 3.5.2 is a typical model for a small information business. Note

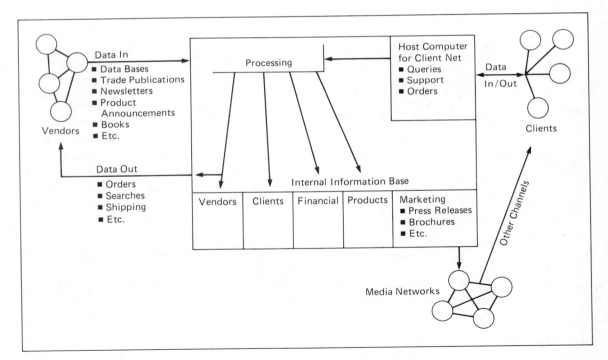

FIGURE 3.5.2. The Info Depot

that at many points in this system, small computers can be used. The traditional approach was to take the most costly function (say, accounting by hand) and automate it using a computer. The current approach says to put a micro at that point in the system where a function is needed. The advantage to the micro approach is that the total investment in a completely automated office need not be made all at once. Micros can be added one at a time, then, later when growth and funds permit, integrated into a LAN [4.16] to bring it all together.

3.5.3   Office Automation

As stated earlier, every segment of our society and economy is filtering the new information technologies through the sieves of previous experience. Thus "electronic publishing" and "office automation."

The functions of traditional "offices" are moving to chips. "Office automation" views the Information Age as chips—microprocessors, terminals, modems—moving into offices. The real transformation, however, is that the functions formerly needing centralized places for the storage of paper-based information that a group of "clerks, secretaries, and executives" could access are now performable anywhere, using distributed terminals with access twenty-four hours a day.

Up-and-coming *office automation experts* call your micro a "multifunction integrated workstation" and thus justify their own high fees and try to perpetuate the remaining aura of mystique that surrounds what is pretty ordinary information processing. There is even talk of an "*intelligent* multifunction integrated workstation," able to handle "data entry" (keystrokes from a keyboard or mouse or numeric keypad), "word processing" (creating and formatting text on a screen or on paper), "voice messaging" (recording a human voice in the computer for playback at a later time), "electronic mail" (private messages from colleagues and/ or friends), and videotext (information from data bases, according to office automation (OA) persons) all at once. The argument being put forth is that all these functions must come from the same place, be in the same software/hardware product, and look the same to everyone everywhere at all times.

These functions are seldom all needed at the same time by the same person. In any case, most of them are currently available in easy-to-use off-the-shelf packages, with varying degrees of interchangeability, depending on need. Seldom does the software developer with expertise, say, in text editor development have the same degree of expertise in financial modeling or communications. And products developed by software teams over many years are notoriously expensive and become obsolete pretty fast in one way or another anyway.

According to one survey, most of the personnel managers of the 1,300 top corporations in the United States either had already introduced or were about to introduce micros, word processors, and electronic typewriters into their offices. The rallying cry? Increased productivity! Never mind that no one has developed an acceptable method of measuring productivity among office workers, particularly executives. Text editing and preparation is almost always the first office function to be automated. That's because at least 80 percent of all messages generated in a company are for in-house purposes.

For those who want to "ease" into automation in the office, leaving plenty of time to think of the whole system, a good place to begin is communication. Various surveys and studies have shown that in traditional business organizations, about 90 percent of all paper business documents consist of internal or *intra*organizational messages. Ninety percent of all these sorts of communications consist entirely of alphanumeric characters: letters and numbers, no special symbols. To state it another way, about 80 to 85 percent of all business documents are textual and numeric in nature and will stay within the organization where they were generated.

Another startling finding is that about 75 percent of all phone calls made do not reach the intended recipient on the first try, even in "telephone-efficient" countries such as the United States and Canada.

The conclusion is that in most offices today, the adoption of some form of in-house computer conferencing or other electronically based message system for both internal and external communications would be a cost-effective move. A small-scale pilot project or feasibility study is a prudent way to go while the external technical chaos sorts itself out.

The main problem, then, will be in getting such pilot projects and small-scale systems introduced—and, where introduced, used by the majority of the organization's participants.

3.6 STANDARDS AND STANDARDS ORGANIZATIONS

There has been little, if any, progress toward "network standardization." This is probably as it should be at this stage in the evolution of silicon-based systems. There has been a proliferation of proprietary local networks. Xerox's Ethernet is a good example of a standard that has widespread support in the industry, but is by no means universal.

Within your personal and small-business softspaces, you will probably end up creating some of your own standards for various things, including networks. There are two approaches, both of which, in tandem, should prevent your softspace from either going out of control or becoming hopelessly outdated by advances in technology.

Strategy A: Adopt your own internal standards. As you evaluate new hardware, software, and networking schemes, make sure new additions support your standards.

Strategy B: Hedge your bets. Adopt more than one standard in each area and create "hierarchies" of standards. For example, you might want to use more than one programming language. Choose two, or more, for in-house use.

A third approach, requiring more dedication to the technical, is to join one or more of the national and international bodies that actually create standards, and make your voice heard—as an *end user,* not an "expert"—concerning the various standards that are being made. This is as much a political process as a technical one, and, especially on the international level, a most intriguing study all its own.

_____ 3.7   POLITICS AND TELEPOLIS

The Age of Complexity: massive systems, poorly understood and even more poorly managed, confront the individual and the community. In the United States, the "democratic process" and "the marketplace" together are supposed, somehow, to deal with these macrosystems. The criminal justice system, drugs, foreign policy, energy, environment, the economy: all of these are names for systems and systems behaviors that we lump together in one big semantic muddle.

There is a tendency for us to expect our political systems—local, state, and national governments, specifically—to deal with all these macrosystems on our behalf. "Fix it, please." Participation on the part of John and Jane Q. Citizen is confined to voting in elections held at lengthy intervals. The percentage of citizens so participating is dismally low year after year, even in presidential election years. When you're not being primed to vote, you are seen merely as a part of a "public" whose opinions are manipulable by well-placed "leaks" and other newspaper and television stories.

Meantime, the slippage or lag time between *cause,* or decision, and *effect* produces outcomes (events) of large magnitude that impact on many individuals adversely, sometimes weeks or months after a particular bureaucratic or congressional decision has been made. For example, many decisions made by the Federal Communications Commission (FCC) and Congress concerning telecommunications have had and will continue to have direct, tangible effects on the quality and cost of your communications. In the case of the deregulation of AT&T, some of these effects (higher rates) didn't occur until well after the original debates and decisions took place.

The catalog can be continued: regulatory agencies that become parts of the systems they are meant to regulate, to the detriment of any real regulation; unions that resist innovation in order to protect the obsolete jobs of their workers, resulting in competitive disadvantage in world markets, resulting in the loss of jobs for the union's members; and so on and so on, down the list of contemporary anxieties.

Using the arts of the netweaver, it is now possible, perhaps, to re-create the form of democracy found in the ancient Greek city-state, the New England town meeting, and the Israeli kibbutz. Every citizen *could* have the opportunity to participate directly in the political process on specific levels. Using existing technology, millions of people could become parts of a *telepolis,* engaging in dialog with one another and with representatives concerning specific issues and problems.

Such an "electronic democracy" has been field-tested to a great extent by various academic researchers, corporations, and other communities throughout the United States and Canada. Early experiments used conference voice telephone calls, linking people at the same time from different locations in an electronic conference.

Current netweaving technologies, of course, are orders of magnitude better than just the voice telephone. Small networks (thirty or less) could fully discuss and examine issues, with input from appropriate resource persons, and then vote on them. The results of these votes are then sent to the appropriate systems and subsystems, composed of representatives from each of the smaller networks, existing controlling agencies, Congress, and so on.

The networks composing the "New Right" have been almost entirely organized using computers and broadcast communications. Clearly, the Information Age is also the Age of Telepolis. Assuming that telepolis is doable and desirable, where does the novice netweaver go from here? There are questions to be answered, to be sure.

- What network organizing strategies do we use?
- How do we examine alternative scenarios and choose the ones we wish to make real?
- How do we make changes less threatening, both to current vested interests and power holders and to the currently disenfranchised and powerless?
- How do we make sure that increased power is also shared power?
- How do we design checks and balances into our networks, similar to those set forth in the Constitution?
- How do we make the designers *and* users of the emerging telemedia *aware* of such needs and questions in the first place?

These questions have no easy answers today. Netweavers will help answer them.

Here are some of the currently unresolved telepolitical issues of the immediate-term future for you to consider:

- Deregulation issues in the telecommunications industry. The arenas currently are Congress and the FCC, to a lesser extent trade and consumer organizations.

- The issue of *data encryption*. Currently, *there are no secure systems* for the encoding and transmission of data. By secure, I mean safe from *all* forms of eavesdropping and bootlegging, including that performed in the name of "national security." This issue is connected with the overall one of computer crime and data security. Focusing on teenage hackers is misplaced: we should be watching the biggest and most powerful computers for bigger and more powerful abuses of citizens and organizations.

- The issue(s) of national and international standards for telecommunications: signal definition, protocols, and so on.

- The issue of the flow of data between countries, sometimes called *transnational data flow.*

- The issue of automation and its impact on the distribution of "consume-ability" throughout societies. People who are put out of jobs by robots that cut the costs of mass production have no money to buy the fruits of said mass production. In the long run, if not the short, this is a grand systems dilemma.

- The issue of *information as education,* and education as information. As expertise, knowledge, and information burst out of the university and into the community, how will we be able to continue to justify the baby-sitting that often passes for mass education?

- The issue of the domestic monitoring of message traffic on telecommunications channels. This is related to the problem of getting a secure encryption method in place for everyone.

- The issue of protecting and enhancing the Freedom of Information Act. This essential piece of Information Age legislation has come under increasing attack since 1980.

- The issues surrounding the problem(s) involved in upgrading the entire telephone/telecommunications networks on the whole planet.

- The issues surrounding the protection of the rights of creators of new information (artists, writers, moviemakers, software designers) in ways that do not restrict the rights of information users (sometimes the same people). Generally, the whole area of information economics is fraught with contradiction, strangeness, conflict, and paradox.

The common theme in these issues is information: Who has it, who gets it, who gets to control it, if anyone? How should it be classified? Who decides and for what benefit?

If all kinds of information can be encoded in digital data form, then it follows that a single, integrated system for handling digital data on a planetary basis is the ultimate end goal. That this can be accomplished, technologically and scientifically, there is no question. All the issues and arguments have to do with implementation. If the single form of planetary information exchange is kept in mind as a goal, then working backward, from that single form, is a useful way to approach the issues. A netweaver's approach—holistic, cybernetic—to the politics of technology might be what the planet needs at the moment.

3.7.1 Banking and Credit

In many scenarios created by futurists, economists, and telecommunications experts, the current functions of banking and credit are pretty much taken for granted. Common features of these scenarios assume less need for travel because we are all linked together. We bank and shop from our homes or offices. In the future, we carry little or no cash, thus we are not tempting targets for street muggers. Every establishment accepts our electronic ID and bills our bank credit accounts directly from *their* handy, dandy terminals.

Unless market forces are quick and powerful, old technologies—in this case the technologies associated with banking—have a momentum that can keep them going long after they are obsolete. So far, the main uses for information technologies in financial institutions have been to streamline current functions and to offload current functions onto the bank customers, resulting in lines forming in the street for automated teller machines (ATMs) rather than in the bank for human tellers. Current experiments under way in California and elsewhere allowing microcomputer network access to bank services offer little that is new or even useful to the average "consumer."

The replacement of gold by paper money, then the replacement of money by checks, were both "revolutionary" in their day. It made sense, in the days of gold and paper, to restrict access to and manipulation of these to special, secure institutions that took the burden of safekeeping off individuals. Banks also made possible the streamlining of investment in new technologies and the efficient transfer of goods and services between people.

What seem to be on the horizon are vast computer files held by banks, containing full details of all our transactions. The consequences of this may be fine for banks, but disastrous for the rest of us, from both libertarian and financial points of view. We need to look more closely, in the Information Age, at the very *idea* of money as a form of information. We need to look closely at more alternatives to money and its now-accepted equivalents (credit cards, letters of credit, checks, and so on).

If money becomes nothing more than a digital record of earning, spending, and investment, then it is highly likely that the function of *banks* will also begin to spread electronically to institutions not normally thought of as part of the "financial community."

Large retail department stores, manufacturers of products, and service businesses are already beginning to handle more of their own credit and other financial services. With integrated electronic networks, it may be possible to revamp and sidestep our entire antiquated monetary system. A worldwide credit system based on energy units rather than gold units or national currency units may hold the answer to the problem of "monopoly" on financial information and transactions.

3.7.2 Macrosystem Conflicts and Competition

Mother Bell prior to 1984 has been compared to Mother Church before the Protestant Reformation. The information environment in which netweavers must operate since the breakup and divestiture of AT&T is at once exciting and complex, chaotic and patterned. The challenge for the netweaver is to discern/create new patterns in telecommunications in the midst of swirling options. Future shock has metamorphosed into option shock: the user/consumer of electronics, communications, networks, and media has never been faced with this many choices.

In 1913, AT&T got its monopoly from the U.S. government because it was believed that there existed at that time a "natural monopoly" in telephone communications. The benefits to everyone of having a single telephone communications supplier outweighed the benefits of a "free market." With monopoly came regulation by the government. Some analysts believed then and still believe now that we would never have become a "wired nation" as fast as we did (in under one hundred years) if this tradeoff between the ideology of capitalism and the economics of scale hadn't happened.

Since 1913, bureaucrats have become accustomed to "calling the phone company" when their own companies or agencies had telephone communications needs of any kind. This was sometimes called end-to-end service because Ma Bell could take care of both ends of the communications link and everything in between, no matter where the "ends" were located. Now, with micros proliferating on office desks, and with vendors of hardware, software, network services, and phone services coming out of the woodwork, the process of selection, coordination, and implementation of telecommunications softspaces requires more than a single phone call to the account rep of a single company.

Divestiture and deregulation of AT&T are not the results of Reaganomics, but the end result of an inexorable technological push. What made technological and thus economic sense in 1913 no longer makes sense in 1984, an era of microwaves and satellites and LSIs and systems analysis. Besides, without AT&T, IBM would have had a less significant challenge in its growing hegemony over data processing. Unraveling all the arguments for and against these changes is hard enough. But one significant ironic note has emerged: Ma Bell's 1983 accounting was based on 1893 railroad accounting. The system demonstrated systemantics behavior. It is now more out of control than ever.

It can be both fun and educational to enter the Information Age in 1984, especially if you enter it without the baggage of past categories such as "data processing" or "management information systems" and "the telephone company" holding you back. It can also be a great advantage if you or your company or organization is starting from scratch, without the baggage of internal departments that no longer make organizational sense in terms of the new information environment in which they have to operate. Between now (1984) and 1986 or 1987, trying to budget a telecommunications-related project will be like trying to chart a course through an asteroid cluster. As of this writing, for example, even the billing *procedures* for the new entities created by the breakup were not known for certain, much less the rates.

Tip for netweavers: The upheavals in telecommunications over the next few years will create a demand for a new breed of consultant and consulting firms. Companies classified as "small"—under $500 million—will need cost-effective means to sort through their communications options, both internally and externally. Smaller regional firms familiar with the global information environment as well as the idiosyncracies of their regional telephone company will find themselves well placed to provide guidance to these smaller companies, as well as to individuals and startups.

In addition to what they're already doing, each regional telephone company and AT&T itself will be able to enter into emerging communications fields. AT&T, for example, will be able to offer "terminal equipment" to end users: that includes everything from telephones with chips in them to microcomputers that have voice phone jacks built in. AT&T will continue to operate its "Phone Stores," and will sell other manufacturer's equipment through this more or less retail chain. It will be able to get its fingers into cable television, including cable digital services. Even after the breakup, AT&T remains huge, with some 400,000 employees and $34 billion in assets (only a tad bigger than IBM). The company will have to set up separate subsidiaries to enter other competitive markets.

Competition between macrosystems means conflict between classes of individuals (MIS managers versus users; home telephone users versus business telephone users). It also means cybernetic "wars" between the macrosystems. The regional phone companies will compete with one another and with AT&T for some kinds of your business. AT&T will go head to head with IBM, both will continue to go head to head with Congress and the FCC and even end users. Netweavers may find themselves more rather than less involved in political processes. Local rates, access charges, and national tariffs are all influenced by politics. AT&T has some advantage here, since it wields enormous clout and coopts through communication (why, didn't you know? AT&T *invented* the Information Age, or so its corporate pundits would have us believe) better than the national media networks *or* IBM, which is no slouch itself in the world of Washington lobbyists.

Already, end-user lobbies are beginning to form to begin to counteract some of the power of the largest telecommunications companies. Residential phone users' groups may well find themselves pitted against business phone users' groups. Where it will all end up or go in the long run is anyone's best guess. Netweavers need to pay some attention, however, to developments. Otherwise, the netweaver will be just another one of the sheep herded into the fleecing corral by the top ten telecommunications companies.

_____ 3.8   GLOBAL ACCESS

Earlier [1.6] I listed the "critical subsystems" identified by Miller that make up all living systems, from the level of the cell to the level of social systems. Some of the properties listed are shared by or enhanced with micros. In particular, the "channel and net" can be traced through the various levels or holons of the global system. This can help the netweaver understand upper-level nervous systems, beginning with the "wiring" of Gaia's brain.

By way of reminder, Miller defines the *channel and net*work as "the subsystem composed of a single route in physical space, or multiple interconnected routes, by which markers bearing information are transmitted to all parts of the system."

The Cell: A chemical information network joins components throughout your cells. RNA signals cytoplasmic organelles. Inducers and energizers move to sites where they control metabolic processes. Hormones from your various gland cells move from the place where they're made to a place where they're transported out of the cell, or into storage between cells. Neurotransmitter chemicals, which can be used as markers to carry signals across synapses, move down axons to cell terminals. Many of these molecular substances move in precise spatial and temporal relationship to one another (frequencies).

The Organ: Control signals from your central nervous system flow to your organs and regulate their action. Signals from your organs are fed back into your CNS.

The Organism: There are two nets to consider at this level: the endocrine and the neural. Information in the endocrine network is in the form of "broadcast" signals. Your entire body gets the message, even though only target tissues may be able to decode and use the information. Neurons, on the other hand, deliver chemical messages—through neurotransmitter chemicals—quickly over a distance to a precise location. This network is "dedicated" and "private" to specific muscles, receptors, and organs. The CNS network is more efficient than the endocrine system, but at the level of your perceptions and actions, the two work together and influence one another.

The Group: Each group member belongs to the group channel and net subsystem. Various sorts of physical channels can connect the group participants. Speech, music, and expressive sounds are carried on the audio channel. Written, kinesic (body language), and facial gesture information is carried on the visual band. To a lesser degree, body spacing and smell (pheremones) also carry information in/through/about the group.

The Organization: Organizational channels and nets connect all the subsidiary groups (departments, task forces, etc.). The communications structure of organizations may be composed of a large number of component systems, each of which may have its own channel and net subsystem, linked in various ways to the whole organization. Each message transmitted on organizational channels may be either explicitly or implicitly addressed to the components that are to receive it. The messages may be transmitted one to one, one to many, one to all, or to whom it may concern. There are other modes as well, each of which will have an influence on the end effect of the message itself. *In other words, choice of the mode of communication is itself a metamessage in social systems.* For example, a message sent between heads of state may be sent publicly, formally through diplomats, or privately through a friend or secret courier.

The Society: Mail services, national telephone services, telegraph, radio and television, print news media, magazines, scientific and technical journals, distributors of books, films, tapes, and records, display advertising companies, lecture bureaus, actors' and musicians' unions, art galleries and museums, public meetings and rallies, demonstrations, private conferences, workshops, symposia, conventions and meetings that include representatives of a number of organizations—all of these and more are part of the channel and net subsystems of U.S. society. Some or all of these are found in other countries as well. There are deciders, censors, and filters in these subsystems. Monetary information currently is controlled almost solely by banks, stock markets, commodities exchanges, insurance companies, and other financial institutions.

The Planet: Since all of the previous levels are interpenetrating, multiconnected with feedback loops of all kinds, Gaia's a synergy of all the previous levels of communication. Because of this interconnectedness and synergy, ill health at any level can have a greater or lesser effect on the global level. Global organizations have emerged to facilitate coordination between societies and nations. Transnational organizations themselves function to coordinate global economic and political events. These include (by no means an all-inclusive list) the United Nations, United Press International, Reuters, summit conferences between heads of state, satellite communications networks, and the Roman Catholic church and other religious organizations.

In the 1960s and '70s, information technology and space exploration presented us with the technical potential to develop a communications network capable of linking every member of every "separate" society with every other living person on the planet, regardless of location. This aggregate and emerging interconnection is the substrate or infrastructure of Gaia's brain. The extent to which it emerges, and how fast it emerges, will depend on actions and behaviors of other parts of the global system.

The computer—particularly the microcomputer—can be employed as a component of the communications channel and net of *all* subsystems, from the cellular to the global. This makes Gaia a kind of evolving "super cyborg" moving with great speed and capacity into an organism that is qualitatively and quantitatively different from anything that has come along in recorded human history.

3.8.1 New World Information Order

Assume that nations are relatively "information rich" or "information poor." Assume, further, that one correlation of information wealth is the number of telephones per one hundred citizens and the number of calls made per year per person. Using these assumptions, nations can be ranked from rich to poor, as follows.

```
RICHER                  <————————————>          POORER
United States
  Sweden
   Switzerland
    Canada
     Denmark
      New Zealand
       Japan
        Netherlands
         Finland
          Britain
           Australia
            West Germany
             Norway
              France
               Austria
                Belgium
                 Italy
                  Hong Kong
                   Spain
                    Greece
                     Czechoslovakia
                      East Germany
```

The fastest-expanding telephone-use countries include Korea, Brazil, Mexico, and Ireland. A country's telephone density, in fact, is about as good an economic indicator as most others, including its rate of GNP growth. Telecommunications as a sector for investment has historically had a low priority in poorer countries and in international organizations purporting to have the interests of the "Third World" and "emerging nations" at heart.

However, it may not be a wise idea to begin with the assumption that a nation is information-poor. The assumption may be due to chauvinism based on cultural factors. Every nation is information-rich if you include, in its information data base, information contained in the culture but not yet put into digital form. The principles of bootstrapping and leapfrogging thus suggest that one good way for a developing country to enter the Information Age with information commodities of its own would be to develop its indigenous information base. This could include:

- Information about itself, its topography, regional characteristics, culture(s), and history.
- Information about its resources: tangible and intangible.
- Information about its people: demographics, census data, and so on.

Rather than assume that a developing nation is an information have-not, a more productive assumption for the netweaver is that its existing systems and subsystems contain "latent information" that could be profitably "mined" and made available to the rest of the world. This assumption about latent information also serves groups and individuals. An inventory of such latent resources should be made part of any systems analysis at whatever level.

Telephones are seldom, if ever, mentioned by UNESCO, a United Nations organization concerned, on behalf of developing nations, with a New World Information Order (NWIO; originally New International Information Order, or NIIO). Such nations want to acquire the information and communications capabilities enjoyed by the industrialized countries. Current information gatekeepers in the United States and western Europe (journalists, editors, news directors, the U.S. Information Agency, CIA, AT&T, GM, program directors, advertisers) emphasize the free flow of information across national borders.

Specifically, developing countries want to end U.S. domination of *all* the media. This includes content: news, publications, movies, television, micro software; and conduits: the use of radio frequencies and satellite orbit slots. Above all, developing nations are concerned about the use of remote sensing satellites over their territories, and of direct broadcast satellites (DBS) *into* their territories, both areas in which claims for "free flow of information" tend to obscure the cybernetic potential of these new technologies. What is at stake is not a reporter's or journalist's freedom to report the "news" about a country, but the function of that same "news" as piped back *into* the country by media dominated by Western industrialized outlooks.

3.8.2   Satellites, Space, and DBS

Developing nations need not go the route of the United States in implementing their own communications and information systems. Figure 3.8.2 shows two means of "leapfrogging" the need for land-based copper wire lines (most of the U.S. phone system). Earth-based microwave stations are required only every thirty miles or so on the earth's surface. Using geosynchronous communications satellites, points on the earth that are 22,000 to 26,000 miles apart can be connected. If the satellite is powerful enough, centralized earth stations aren't needed. The satellite can broadcast *directly* to individual stations. DBS technology combined with digital micro communications and fiber optics could enable any country to leapfrog the United States in information capability.

3.8.3   Video Revolt

In Belize in the early 1980s, local entrepreneurs, using currently available earth stations to pick up signals from U.S. satellites, bypassed the na-

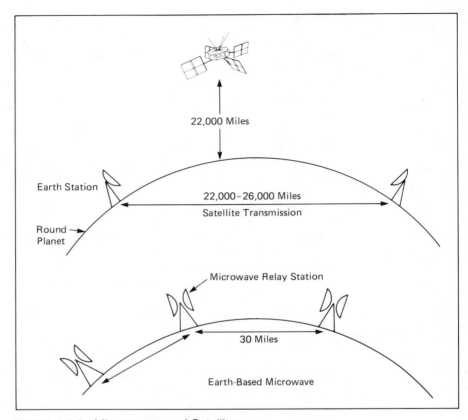

FIGURE 3.8.2. Microwaves and Satellites

tional government and began to provide videotapes of U.S. shows to the locals. These info-privateers stepped in to provide what their government had promised but had never delivered: a television network. The next step was local cable links. Despite what national governments may say about "U.S. domination of the media," it seems that local populations *want to see* the latest Home Box Office offering. It is unlikely that any national government, in an age of mass-produced audio and video cassette recorders, will be able to contain the video revolt, especially when poor nations can become richer as a result of new information technology. The New World Information Order is being built, all right, from the ground up, the grass roots out.

3.8.4   The French

France is employing the leapfrog strategy to its economy, moving more quickly from industry to information. Japan, of course, is well known for its leapfrog over the technology of the West. Some of the best theoretical work of the Information Age about the Information Age has come from these two countries. French analysts Simon Nora and Alain Minc, writing in *The Computerization of Society,* suggest that "a technological revolution may simultaneously create a crisis and the means of overcoming it, as was the case with the coming of the steam engine, the railroads, and electricity." They urge the netweaver, however, to resist the expectation that micros are going to overturn the power hierarchies of our planet overnight. Traditions all over the world, including those in the Second-going-on-Third Wave countries, stand in the way of the initiative and adaptability required by the Information Age. As they point out, telematics can facilitate the coming of the new, but it cannot construct it on its own initiative. For that, dear netweaver, you are required.

Nora and Minc make the best case for the building and controlling of data banks on the part of nations, organizations, and individuals. Information, as they so eloquently point out, is inseparable from its organization and its mode of storage. When you access someone else's data bank, you are also caged by that someone else's *organization of* that data bank. Leaving to others—whether "America," "the government," "IBM," "the Source," and so on—the responsibility for organizing our collective memory, and remaining content to plumb it, is to "accept a form of cultural alienation."

Data banks created by the individuals and communities that use them will become the richest resources of those communities, regardless of what some other country or company tries to sell them.

Finally, if the foregoing isn't enough hope for the experienced or budding netweaver, the French press home the point: "Information that only teaches technical solutions, that lists facts without putting them into a perspective and without structuring them into a coherent project, and, on the other hand, information that proclaims ideals without inserting them into the practical development of society will increasingly be regarded as pseudo-information." Amen.

_____ 3.9   NETWEAVING

An electrical energy grid is in place and growing on the surface of our planet. Helping to weave networks of human communication is an important evolutionary role with implications for people and for the biosphere in which we exist, without which we *cannot* exist. Netweaving begins with the realization that *the grid is not static.* The game is not over; it has hardly begun. Our species has been left a legacy by thousands of cultures spanning thousands of years. Some of that legacy, surely, is poisonous and detrimental to life on earth. But the larger proportion is priceless, worth preserving and extending as we move out to the stars. The choices concerning what to save and what to leave behind are individual choices: yours and mine.

Consider what has been accomplished in the last fifty years in electronics alone. Then consider the insights we have gained from new sciences: our very *awareness* of problems such as environmental damage and population growth relies on scientific and technical advances of the last thirty years or so. Now look ahead: reflect on the things that need accomplishing in the next ten, twenty, or fifty years. Much depends on whether or not we are successful in weaving together hundreds of millions of people with millions of computers in a combined intelligence that can get us through these turbulent times. Paradoxically, it cannot happen at all without *you,* netweaver.

Oh, it may *seem* as though there is more to learn than you have lifetime to learn. It may *seem* as though there is no place to begin, no place to finish. All and everything may *seem,* at times, overwhelmingly hopeless and futile and without a future. It may *seem* that the juggernaut of history is taking you where you don't want to be. But if you link with just one other person in a spirit of hope and love, if you simply *begin,* entropy can be held at bay, if not reduced or finally eliminated. We have the means, everything at our disposal. Reach. Do. Continue. New opportunities now exist to make your voice heard and seen.

3.9.1　Personal Learning Networks

Harriet decided that she wanted to learn more about microcomputers. Naturally, she began by informing her friends of her intention. Some of them already had computers. One or two were already versed in linking up with others in the infosphere. She approached her self-assigned learning task with a diligence typical of her Virgo-ish character. Her data gathering began with friends. Friends pointed out books, gave her articles to read, and were eager to tell her about the virtues of *their* particular brand of micro.

She visited newsstands and scanned the many magazines spawned by micros. She was careful to buy only those that contained information *she* considered useful to her in the task at hand. She found out that her city had several local networks of micro users who clustered around local BBSs [4.14.6]. After doing as much research as she could using the print media, she was ready to make a move.

She made a decision to buy some equipment. Did she go out and buy the latest micro on the market? No. She bought a *used* Vic 20 and a modem—combined cost, less than $150—and had an extra phone line installed in her apartment. Using her television set and the telephone and her Vic 20 and the modem, she began "visiting" her community networks. She spent a lot of time browsing and reading the hundreds of messages being passed around. Many of these messages provided her with further insights into information technology and gave her more ideas about where *she* wanted to go with her personal softspace. She took notes copiously, and after four months of data gathering on her local networks, she was ready to make her next major move.

She bought a printer. (A low-cost, dot-matrix printer—less than $200.) Why? Because there was a lot of information on the local BBS and conferencing networks that was valuable to her, and extensive enough to warrant having her own printed copy. There are limits to manual note-taking, after all. She now had enough of a basic foundation in micros and telecommunications to be able to begin participating in the conversations she had, up until then, only been monitoring. Often, her open questions unleashed a flood of new data from her network community.

She could have stopped there. After all, she had a successful career as a visiting nurse. Her networking and computer-related activities could have remained at the level of an avocation, like having a CB radio. But she didn't.

3.9.2   Remote Supervision/Cooperation Networks

Looking around her community, and assessing her professional associations, Harriet uncovered some needs that were going unmet by any of the existing health care services. Information about available services was badly needed by those the services were intended to help. While there were plenty of health care organizations, family care services, and the like, no one had made an inventory of these services. What's more, people who needed special services, such as at-home physical therapy, weren't able to find out about them easily. The health care network functioned informally.

Using a combination of her professional affiliations and her newly found electronic community, Harriet put together her first network of five people. All five had micros or access to micros. This first netweaving project became a Community Health Care Task Force. Its goal was to put together a comprehensive inventory and directory of health care services in her county. The task force also did a feasibility study to explore the possibility of putting the resulting data base on-line, for direct access by anyone who needed health care services.

As facilitator of the Community Health Care Task Force, Harriet found herself performing management functions for the group. Much of the work of the Task Force took place on-line. Eventually, Harriet and her task force decided they needed their own micro conferencing system to streamline and expand their work. They "formalized" their network by forming a nonprofit corporation to serve as the legally recognized entity hosting their network. The combined resources of the task force, plus their willingness to organize themselves more formally, resulted in support for their project from a local foundation. The problem of a budget for their own computer network was largely solved at that point.

Eventually, Harriet found herself working more and more on the networks. There were still plenty of face-to-face meetings to attend, but those necessary meetings were more pleasant and much more productive as a result of all the preliminary work that was done using the networks and the combined information handling power of the task force.

3.9.3 Family Networks

Bill was the only member of his family who had a microcomputer and a modem. He was enthusiastic about the technology but skeptical about ever being able to involve the rest of his far-flung clan in his enthusiasms. Then his older brother bought a micro for his mother, who was a part-time real estate broker. The rationale was that the micro would help Mom in her business. Aha! thought Bill. We have the makings of a network here. He persuaded the rest of his siblings to buy Mom a modem for her new micro. Then he set up his own micro on his phone line during his work hours and late evenings, when he wasn't expecting voice phone calls. He showed his mother how to access the BBS program he'd installed, how to leave messages and read messages. He wasn't sure if or how the system would be used. He and his mother were only a forty-minute commute away from each other.

His mother surprised him totally. She became so enthusiastic about this means of communicating with her son that she began to show the system and its capabilities to the rest of her family every time they visited. Soon, another micro was added to the network by a sister who lives in Des Moines. Before they knew it, the family had its own network, sharing gossip and recipes, technical help and business information, and just plain fun communicating with one another. It was no substitute for their annual reunions, but with a kinship network of over 380 people scattered over thirty-two states, their micro network created a channel for them that was used much more than paper mail and cost less than voice phone calls to maintain. It took three and a half years before most everyone was using the network, and in that time it had pulled the family through at least one crisis: Aunt Mary and Uncle John had their home wiped out in a flood. The network rallied to help them.

3.9.4 Community Soapbox Networks

In most communities in the United States there are now people who are fully "wired into the networks." Creating local consensus-making networks is now a matter of, first, identifying this micro-constituency. By contacting civic organizations, local government agencies, micro hardware and software sales organizations, including retailers, microcomputer clubs, and one's own friendship network, the seeds of a network-based community forum can be collected and nurtured. Both organizations and individuals can be classified according to "processing power." What kinds of computers are there? How much storage is available? Which softspaces are willing to become "host nodes" for the community?

With the information assets of the community identified and cataloged, the next phase is to begin an ongoing on-line meeting. The first goal of the on-line meeting is to identify community concerns and needs. The process of inventorying is also ongoing, so that as new people find out about the network, the information assets they're willing to bring to the network can be added.

A community newsletter, generated in part from on-line information, is used to contact and bring together people in the area who do not yet have access to the on-line forum. One of the micros in the network—preferably *not* the one that is hosting the on-line activities—is used to maintain and print out a mailing list.

3.9.5   Consultants and Consultants' Networks

It is now possible to become an on-line consultant and to use networks to facilitate one's consulting activities. It is also possible to obtain a college degree wholly or in part using telecommunications networks as an adjunct to the learning process. Consultants in *any* field, at the very least should set up their own "information magnet." Any small micro—Apple, IBM PC, TRS-80—can be used to host an ongoing conference and electronic journal for specialists in a consultant's area. This can be a geographical *and* informational area.

3.9.6   Special-Interest Networks

A Special-Interest Network (SPIN) is any network dedicated to a specific topic or set of topics. A SPIN can be organized in many different ways, depending on whether the network is to use existing information or generate its own information on the basis of user input, or a combination of the two. The creation of an on-line newsletter for a SPIN is ridiculously easy: off-the-shelf hardware combined with off-the-shelf software and a phone line will do it. As with any network, it doesn't really matter how large the system is in the beginning. Certainly the network will outgrow the initial installation if it's small to begin with.

3.9.7   Distributed Network Organization Development

The biggest obstacle to be overcome in developing a network-based organization is the tendency to a *directive* style of leadership on the part of its organizer(s). The initial tendency on the part of someone newly interested in networks and telecommunications as an organizing tool is to abandon all power and leadership in favor of "letting the network do it." It quickly becomes apparent that the network is powerless to coordinate its efforts or to get off the ground altogether. One reaction to this powerless state of affairs is to try to exert directive control of the network. If the person(s) so attempting also happen to control one of the network's important computing resources (say the host computer for the net's communications) then potential participants withdraw their support.

If the network is one intended to create income for its members, the problem of directive leadership versus autonomy can become severe. An organizer can adopt OD methods from the very start or can incorporate them into an organization or network that already exists. The advantages to the organization are several:

- The OD approach gets people into communication with *themselves.*
- The OD approach provides needed *feedback* to individuals so that the impact of their behavior(s) on the group can become conscious.
- The approach fosters a *caring climate,* which in turn fosters interpersonal trust through which feelings, perceptions, and ideas may be shared without internal fear filtering the output.
- It improves overall communications skills.
- It helps people define what they want: for themselves, for the organization, for the future.
- It helps people recognize the systems complexities involved in the interplay of *their own* needs, desires, and goals with those of others.
- It helps in conflict resolution, providing alternatives to fighting, compromising, or quitting.
- It helps bring group goals and individual goals into resonance.

A manager, leader, business owner/operator who wishes to employ OD techniques within the network structure will resolve to:

- Increase the sense of *autonomy and* the sense of *interdependence* of the individuals in the net;
- Increase participants' *understanding* of how they affect both their own destiny and the destiny of the net;
- *Build confidence* in people's ability to affect their own futures and the future of the net.

Individuals in the network cannot effectively perform their tasks without information about the relationship of those tasks to other parts of the network and the net as a whole. Decisionmakers in the network cannot make effective decisions if denied information stored in the biocomputers of other members of the net. No part of the network can implement decisions that seem nonsensical to them. By its overall behavior in the sharing of information, a network communicates to its members that it trusts them and respects them. This enhances the motivation, on the part of individual nodes in the network, to remain loyal to the network and its objectives.

3.9.8   Personal Global Networks

It is now possible for individuals to deliberately create global networks of other individuals for specific purposes: personal, political, social, economic, cultural, educational. Doubtless, as these networks take shape, specific information about strategies and tactics, uses and abuses, methods that work and don't work in different cultures in different parts of the globe, will begin to emerge.

As you enter the global community and help to shape it, bear in mind that what you will be learning, in all likelihood, will be unique and highly specific. Even though there are guides of sorts, the destination is unclear, the path sometimes confusing. Try to keep your wits about you: Mother Gaia will provide.

4

Building and Weaving

In this section, after a short review of how the current state of affairs came to be, we cover a netweaver's array of tools and techniques. This is where the going gets technical and the jargon flows freely. If you feel you are not adequately prepared, go to [1.0] to arm yourself before returning here.

4.1 MILESTONES

The discovery of the electron came as no surprise to those who were looking for it. This discovery led to electrocution and the light bulb. Thanks to Lee DeForest the light bulb evolved to become the vacuum tube. Originally, vacuum tubes were called valves because of a metaphor comparing the flow of electricity (the "current," measured in amperes) to the flow of water. Modified light bulbs could be made to reduce or increase the flow of electricity through them in predictable and useful ways.

Valves were used to amplify small signals. They did this by using weak signals to regulate stronger, more powerful flows of electricity that could then do other useful things, such as driving a loudspeaker or switching a relay. A trickle, small and insignificant, could thus control the flow of the mighty Mississippi. This gave us television, high fidelity, radar . . . and rudimentary computers.

To build a useful computer with electromechanical relays and vacuum tube valves is quite difficult and expensive. You need a lot of glass. The glass has to be shaped into a lot of light bulbs with hot filaments in them (to supply electrons rather than light). The filaments would burn out, just as they do in today's light bulbs. And they generate heat, so the more you have of them, the hotter their environment becomes. You need *lots* of light bulbs to make even a smallish useful computer. Soon you are spending most of your time just changing light bulbs (probably as part of a whole team on roller skates) and most of your available energy on air conditioning to cool out the environment. Making large computers this way is not only expensive, but just doesn't pay for any but the most important applications in the largest of institutions. Fortunately, there is a better way of using glass (silicon) than making tubes out of it.

4.1.1   Brains

The computers that were built out of vacuum tube valves were large, fragile creatures. They caught the imagination of the popular press of the time, however, which dubbed them "giant brains" and "thinking machines." Nonsense! cried some of the scientists of the era. In order to get a computer to simulate even the teeniest of thoughts of the human brain, they argued, we would have to build a machine the size of:

the Grand Canyon (early mechanical computers)

Yankee Stadium

the Empire State Building (valve-based computers)

a couple of floors of the Empire State Building (early transistors)

a single room

six cubic feet

a desktop (VLSI and cryonic computers)

Scientific debates about intelligent machines center on just what kinds of intelligence or human thoughts are being simulated. For what and for whom are big questions, too, but debated less. The underlying assumption is that intelligent machines are benign and will usher in a golden age for humanity.

4.1.2   Transistors

Gaining a complete understanding of how we got from the Grand Canyon to our desktops would require an extensive detour through modern physics and its relationship to electronics. The development of the transistor and, with it, *solid state electronics* grew out of new basic insights into the nature of the atom and the behavior of its electrons.

The first transistors were built on the surfaces of small germanium slabs. They were expensive, partly because they used solid gold contacts, and almost as fragile then as the tubes they were destined to replace. Subsequent generations of transistors became as common as beach sand because that's what they were eventually made out of, and mostly what we use today to make integrated circuits (chips).

The development of the transistor was fortunate for both the Japanese and the computer, which was trying so hard to be born out of tubes. Soon thousands of transistors were being packed onto single silicon crystal wafers. The Japanese are good at this sort of thing. Hundreds of thousands and millions of circuits on single wafers soon followed. The end is not in sight, although the laws of physics dictate that there *will* be an end to all this packing of circuits into ever-smaller areas. Someday.

This evolution from light bulbs to chips to micros to networks took under fifty years of recorded history. If it were merely a matter of turning sand into analogs of human symbolic behaviors, that would be the end of it.

4.1.3   Chips

The first microprocessor family was announced in June of 1971. Thus Chip was born. The line had to start somewhere, and the first family consisted of all siblings: the four-bit 4004. It began as a custom design, by America's Intel, for a Japanese calculator manufacturer, Busicom. The 4004 chip set was designed by Federico Faggin, who went on to become president of Zilog, Inc. By the end of the seventies, Chip had become a sixteen-bitter, exemplified by Motorola's 68000.

The wide availability of microprocessors—"brains on a chip"—made possible the design of actual, working computers in personal "laboratory" spaces like garages and cellars. In addition, it was only a matter of time before over 250 manufacturers old and new began using Chips of all generations to design and market their own versions of information-handling-if-not-thinking machines. Here it is the mid-1980s and things haven't calmed down yet. They are not likely to, according to those who monitor such things, because before it's anywhere near finished there will be intelligent electronic softspaces almost everywhere. In fact, our whole environment will become the Ultimate Intelligent Softspace.

4.1.4   Chronology

The following is a chronology of events leading to the era of networks:

c. 2600 B.C.	The Chinese invent the abacus. So it remains for a long while.
c. A.D. 1642	Nineteen-year-old computer wizard Blaise Pascal invents a mechanical adding machine he calls "La Pascaline."

1694	Mathematician Leibnitz improves the Pascaline. He designs a mechanical dummy that can multiply, divide, and extract roots.
1835	Charles Babbage, in a memic fit, creates the Analytical Engine, a true forerunner of the computer using punched cards to hold data.
1890	This is the Year of the Great Census Machine. This device, invented by Hans Hollerith, bails out government statisticians who are up to their alligators in census data. Thus is born the *data processor.*
1939	A landmark computer, the Mark I, is built out of mechanical relays.
1946	The first Electronic Numerical Integrator and Computer—ENIAC—is built, using electrical switches instead of electromechanical relays. A valve burns out every seven and a half minutes . . . the smell of frying brains.
1951	This is the year of the first commercially available computers, which go to big government and big business. Artificial intelligence is still in the wings.
1956	For the first time in the history of the Second Wave, white-collar workers outnumber blue-collar workers. The mechanic has become the clerk.
1957	Sputnik is launched. Gaia's nervous system goes global.
1958	The first telecommunications satellite is launched: Signal Communication by Orbiting Relay Equipment (SCORE).
1960	Transistors are introduced to computers. They marry and beget exponentially.
1970	The first large-scale integration (LSI) of circuitry puts hundreds of transistors on a single silicon chip.
1970	The first computer-conferencing network is created.
1971	The first programmable microprocessor hits the commercial marketplace.
1975 (January)	The ALTAIR computer hits the hobbyist market. Many are sold, accelerating further development primarily by spare-time amateurs.

| 1975 (June) | The first storefront retail computer store opens in Santa Monica. |
| 1976–present | The rate of change (acceleration) increases. The Dawn of Tofflerism and the Electronic Cottage. Empires are built on the mass-market micro. |

4.2   ENTRY LEVEL: PERSONAL SOFTSPACES

You, a comfortable reading chair and lamp, and a book or magazine constitute a familiar information system. At least since the invention of the printing press, the delivery of paper to people has been the primary channel into personal softspaces [1.7]. Then came radio. A few more chairs were added to accommodate more of the family or neighbors. Keep the radio in mind as you introduce a television set and, for convenience, a telephone nearby. Are you counting channels? If you now envision a micro in the setting, you suddenly and with great impact add a marvelous gateway with countless channels to your softspace. The informative reach of your softspace is dramatically expanded when you link the telephone and the television through your micro.

That was easy, wasn't it? The softspace is linked to your senses through primarily physical means: paper, speakers, video displays, printers, keyboards, mice, touchscreens, joysticks, bit pads, and on and on. Your physical surroundings, the equipment you use, the storage and access capabilities you command, all constitute your softspace.

A softspace is process-oriented, denoting an experience through time, facilitated or hindered by surrounding physical factors. Your personal softspace may be centered in a corner of your study, with a computer terminal linked to your telephone, and may include a data base in Des Moines or a games network spread out over several cities.

So, even at "entry-level" prices in the $100–$500 range (c. 1984), your personal softspace can give you many mansions'-worth of expanded communications and information access. It should be noted that many companies are now trying to jump into your softspace. *They want access to your information system, if not your nervous system.* They want to be the ones to determine what, how, and where information is delivered to your home or office. People who now only have channels (e.g., cable television companies, FM radio stations) or make receiving devices (e.g., television set manufacturers and cable decoder boxes) or supply information (e.g., newspaper publishers) are coming up with alliances and different schemes for getting you hooked up to their version of the Information Age, a proprietary version at that. Most of these versions have little or nothing to do with your personal microcomputer system. They are essentially *teletext* [4.2.3] services.

Many of these experimental and trial services are based on selling you an "add-on" device for appliances you have in the home already. For instance, some television manufacturers are beginning to place decoders in their sets that will be able to unscramble their scrambled for-pay information. The information itself can be sent along with the regular television signals in the *vertical blanking interval* between picture frames. Again, some television set manufacturers are taking the telephone approach and putting circuitry in their boxes that will connect directly to the phone lines. The mythical end user then dials up information that is displayed on the TV screen.

There are at least eight identifiable classes of activity and information that a fully equipped softspace (i.e., enough local computing power tied into the network[s]) should be able to handle:

Information retrieval: access to remote data bases, news, abstracts, and so on.

Education: access to learning networks, both private and public, leading to enhanced skills or even a degree.

Transactional services: reservations, some banking transactions, purchases, and so on.

Message services: electronic mail, computer conferencing, bulletin boards, public and private "information utilities."

Telecommuting: the ability to do a certain amount of processing-for-pay from one's softspace.

Teleprocessing: access to enhanced computing power and computer programs that can be run on your local micro.

Telemonitoring: the ability to monitor conditions in a home or office such as temperature, illegal entry, the status of appliances or critical equipment, pressure and other sensory information.

Games: access to network-based games for learning, entertainment, and so on.

As you can see, these categories overlap. They will become more arbitrary as time goes on, simply because these separate functions are being integrated with others on a given service or using a particular combination of hardware, software, networks, and outside services to construct your softspace.

4.2.1   A DON'Ts Checklist for Beginning Netweavers

In the short-run future, a lot of manufacturers and would-be entrepreneurs looking to cash in on the Information Transformation are going to be offering a wide variety of combinations of media and media products. Many of these will be one-shot attempts at some proprietary version of videotext or teletext. A lot of them will fall by the wayside. Businesses, especially, should be wary of implementing grandiose LAN schemes and equally wary of vendor-offered services promising to serve all your needs now and forevermore. Any netweaver who wants maximum freedom and softspace flexibility in the long run should definitely heed the following DON'Ts.

- Don't buy services that require you to buy or lease a special black box that can *only* be used to decode and display digital information from a single source. That box will lock you into the source and type of information that will become more limited with the passage of time, not less. In the long run you will spend as much on renting or leasing that box as you would on getting your own terminal, micro, and modem for use with the phone lines.
- Don't wait for digital services to become available on your cable television channel sources. There is plenty of information to access out there and available right now, much of it free or low cost.
- Don't bet on the long-term success of experimental systems such as FM digital broadcasting (using FM radio and yet another "black box" to pick up software programs material, for example) or public radio services or private "video networks." They are not yet sufficiently "user-definable" to be valuable, and may never be.
- The terrestrial telephone systems, linked with the expanding satellite services, will likely be the main conduits into your home or office. After all, when you can link anything to anything (well, *nearly* anything) using the telephone lines, why go with a service that only links you to one supplier or set of suppliers? This same argument, by the way, holds for any mass-oriented "one-stop" information supermarket trying to be everything to everyone.

Figure 4.2.1 is a schematic representation of the difference between a teletext and a micro-based (videotext) network. The components of the latter are covered in the section numbers shown in brackets.

4.2.2   Videotext and Teletext

The generic term for receiving *and sending* textual information on a video screen is *videotext.* (You may also see this term being used without the final *t.* Either spelling is correct. Netweavers will generally avoid confusion, especially among nonspecialists, by including the final *t.*) Videotext and teletext systems are often spoken of in the same breath. The only widely agreed-upon difference between the terms is that teletext, or *broadcast teletext,* as it is also sometimes called, is essentially *one-way* and based on a broadcasting model of communication. Videotext, on

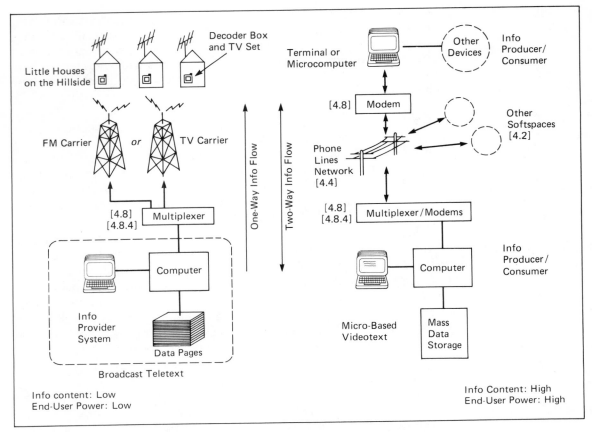

FIGURE 4.2.1. Broadcast Teletext and Micro Videotext

the other hand, is essentially two-way and allows the user to actively process and send information to the central or host node of the videotext network. Videotext experiments and actual, usable systems have varying capabilities and give varying degrees of information-creating and information-processing power to the individual users. Netweavers interested in setting up their own services or accessing the services of others can safely ignore the teletext model.

Info industry projections call for $38 billion by 1990 for paid electronic services (both teletext and videotext). Entertainment now accounts for the majority of revenues in this sector. The home or office microcomputer is the delivery device of choice. Phone company (AT&T) services will probably outweigh all the rest, including various forms of cable (coax, fiber optics). Soon, both text (alphanumeric) and graphic images will be routine. Currently, text-only systems are more common. Graphics and animation, however, are in the works for the very near term future [4.9.5].

As I pointed out earlier, there are at least one hundred active companies (c. 1984) either offering or planning to offer some form of videotext information services to home and business users. These include information providers, equipment and software vendors, and conduit owners.

With teletext systems, all the user can do is take what is offered, when it is offered. Even though teletext systems are limited when compared with videotext systems, it would probably be unwise of the netweaver to rule out teletext completely. Under certain conditions and for certain kinds of information, teletext may be entirely adequate.

4.2.3   Data Communications

All the "-text" systems are variations on the theme of data communications. For the netweaver, data communications means getting micros to talk to larger computers and to each other. The phone lines and microcomputers are the preferred means of accessing the new information resources. The underlying technology of data communications makes videotext and teletext possible. Some understanding of data communications is required before attempting to become a designer or even a skilled user of networks via micros. Data communications is concerned with the specification of modems, types of channels used, protocols and standards, and so on. The phrase itself grew out of the monolithic computer environment of yesteryear and has since broadened in scope to cover many different kinds of digital telecommunications.

Strictly speaking, data communications goes on inside your personal micro all the time. The keyboard is communicating with memory and the CPU, the CPU to the disk drives, memory to the video screen or hardcopy printer, and so on. Some data communications has to take place between the computer and whatever peripheral equipment you have attached to it. Many of the principles of such internal communication can be applied to long-distance data communications.

In spite of all the hoopla about videotext and teletext, in general those touting these special categories of telecommunications are selling obsolete technical devices and don't consider the end user at all, except as a "consumer of information." These media also continue to prop up the obsolete distinction between the information *provider* and the information *consumer*. As micro guru Ted Nelson put it, "This creates two different castes, one bestowing information on the other like gods showering largesse down on mankind, or perhaps farmers scattering garbage to swine."

Teletext and videotext, as they are being written about and promoted today, are mass-oriented data communications using general-purpose machines (computers of all kinds) to simulate the existing broadcast media. There is nothing inherent in the technology that says that the broadcast model is the only valid one, or the only one worth simulating.

The only arguments *for* this model are based on Second Wave—mass production/industrial—economic reasoning.

_____ 4.3   ## THE TELEPHONE SYSTEM(S)

Telephones: there are 500 million of them installed around the world. In theory, any one of them can call any other. Most of us hadn't given much thought to our telephones before, say, 1980 or '81. Around that time, the number of telephone options being advertised for personal and business use began to multiply.

Cordless telephones, smart and brilliant telephones, phones that remember, and phones that answer their own calls now come in a dizzying array. Of course, you may also buy just the plain vanilla phone that Western Electric has been providing for years. (But shop around before you buy the phone that the company has been renting to you all these years. It's reliable, but isn't such a bargain as it once was.) The trend, increasingly, is toward ownership of telephones. And, also a trend,

FIGURE 4.3. Novelty Phone: Kermit the Frogphone
© 1982 Henson Associates, Inc.

telephones are being transformed in both function and form as chips take up residence inside them.

The fact is that over 90 percent of all American homes are wired for telephone service already. With that kind of delivery system in place, is it any wonder that expert observers are betting on the telephone companies to deliver the bulk of digital communications services to Mr. and Ms. User in years to come?

4.3.1 Regulations and Tariffs

A tariff is a list of services and rates to be charged for services. Common carriers—or any other organization, for that matter—wishing to offer communications services in the United States must first submit a tariff to the Federal Communications Commission, the government body charged with regulating all kinds of communications. An organization that has had its tariff approved by the FCC is then issued a license and becomes a common carrier.

4.3.2 Common Carriers

If the primary function of a communications organization is to offer services to the general public, it is called a common carrier. In the beginning, there was only one major common carrier, and its name was Ma Bell (the Telco, AT&T). Because it was a "monopoly in the public interest," the telephone company was heavily regulated by the FCC, which is in charge of all "public-interest communications" issues. Now, however, there are other common carriers. In the spirit of "deregulation," mostly evolving out of changing technologies, it is now possible for other companies to become common carriers.

4.3.3 Value-Added Carriers (VACs) and Specialized Common Carriers (SCCs)

A value-added carrier is a telecommunications service supplier that provides more than just a channel for the transmission of information. VACs are important in data communications because they can supply services that may only be needed under special circumstances. Telenet and TYMNET, for example, are two VACs that provide long-distance message transmission services for computer communications. They process the signal to provide error detection and correction, packet switching [4.4.2], and high-speed transmission and/or storage for users.

"VAC" and "SCC" (specialized common carrier) are sometimes used interchangeably to refer to specific companies or groups of companies involved in supplying communications to end users. A value-added carrier adds something to its service that sets it apart and supplies the needs of some segment of the communicating public. Specialized common carriers do exactly that: they specialize in some kind of added value service. Both VACs and SCCs may be referred to simply as alternative telephone companies.

There are more alternative phone companies with each passing day, and more are on the way. Alternative voice phone services such as Sprint (Southern Pacific Communications) and MCI are examples. TYM-NET and GTE's Telenet are VACs providing packet switched networks and sophisticated message conditioning to computer users (micro and all others).

You may also see such systems referred to as value-added networks (VANs—a potentially confusing acronym). The primary thing for net-weavers to notice about all these silly initials is that they don't tell you much at all about any specific service. Mostly they serve as buzzwords in expensive workshops on data communications and "network management."

4.3.4   Telephone Service Resellers

A "telephone resale service" contracts with AT&T or one of the SCCs for a "dedicated" or "leased" line. The service then resells time on the line(s) to end users. By keeping such dedicated lines constantly busy, and/or by using a computer to route calls over the cheapest available service at a given time, a reseller can make money.

From the netweaver's point of view, using a reseller is much like using MCI or Sprint or one of the other SCCs. There are a lot of them, and more are on the way as satellite transponders and transponder time get cheaper.

4.3.5   Choosing an Alternative Phone Service

Keep a log of your long-distance calls. If you are already using your regular voice line to access other computer-based services, this may not help. It will give you feedback on how much your telehabits are costing, however. Note the time, place, and duration of the call. If you save your phone bills as they are sent to you, you already have a log of this information on hand for analysis.

Rule of thumb: you save most on long-distance calls made in the evening hours. You save most on local calls during the day. Arrange your network accordingly.

Things to check in the alternative service:

Is there a monthly minimum? If so, how much?

Is there an initial deposit required for the service?

Are there any first-minute charges?

What are the geographical limits of the service?

Are there time constraints on the service?

Are there access restrictions based on time of day?

Make sure you get a copy of the rates, locations served, and rules of operation of any service you are considering. If it cannot provide a printed copy of such information, pass 'em up.

Even after signing up for a service, you may discover that there are things about it that you don't like. Don't be afraid to enter the research phase once more and change services until you find the one that meets your needs and requirements.

While many of these alternative services advertise savings of "50 percent or more" on your long-distance bills, you may not realize such savings. Much depends on which cities you call, how big your monthly long-distance bill usually is, and other factors. Surveys by consumers' groups have shown that you may not, in fact, save much at all over AT&T's long-distance service by going to an alternative provider. In the intermediate-term future, AT&T's long-distance rates will be reduced, and some alternative services will probably increase their rates, thus making the prices pretty much equal across the board.

4.3.6   Telephone Usage Checklist

This list should provide both practical and financial help for you.

1. Check your monthly bill to make sure you are getting the services for which you are being billed. If you make a lot of long-distance calls, keep a phone log and compare the log with your bill each month.

2. A phone log kept for a month or two may also show you how to save additional money. Most local phone companies offer two methods of billing for you to choose from. In many cases, the lower "life line" or "economy" or "low use" service is a better bet. This is especially true if you have a separate line installed for your micro communications activities.

3. Only private lines are covered by the FCC ruling allowing you, the customer, to attach your own equipment to the phone lines. If you are using a party line or a pay phone, it's no dice, Charley. However, when you use an acoustic-coupled modem, this ruling does not apply, since you are not really "attaching" anything to the telco's precious little network thereby.

4. Installing your own wiring in a new house or apartment that has never been wired for phones before may not save you any money. Comparison-shop first. On the other hand, installing your own extension line and jack may well be more economical.

5. As of this writing, only thirty-one states allow you to wire your own home for telephone service. Other states will probably soon follow suit. The thirty-one where it's okay are the following:

Arizona	Massachusetts	Ohio (Cinci Bell only)
California	Maryland	Oklahoma
Connecticut	Minnesota	Oregon
Colorado	Mississippi	South Carolina
Florida	Missouri	Tennessee
Georgia	Nevada	Texas
Idaho	New Hampshire	Washington
Illinois	New Jersey	West Virginia
Indiana	New York	Wisconsin
Kansas	North Carolina	
Kentucky	North Dakota	

6. If you or a client is unable to use the phone directory to look up numbers, either because of visual impairment or other special circumstances, then the phone company cannot charge for your calls to directory assistance (DA). Contact your local business office for details.

4.3.7 Touch Tone or Pulse Dial Equipment

In order to access some of the alternative long-distance services, you will need to use a telephone that can generate "true" Touch Tone signals, or a touch tone generator, available separately. Touch tone service on a phone may cost more money than the pulse dial type. Some telephones *look* like Touch Tone phones, complete with buttons, but they may only be able to generate pulses when the buttons are pressed. The pulses are made by opening and closing the circuit at regular intervals. The pulses can then be detected by the switching circuits at the central exchange. A Touch Tone phone uses pairs of audible frequencies to operate digital switches in the phone company's switching computers.

4.3.8 Telephone Equipment Checklist

The following is a list of things to consider when buying telephone equipment:

- Buying your own phone not only may be cheaper than renting one from your local phone company, but may also give you additional convenience features as well.

- Sometimes an investment in a wireless phone is better than having extra wiring and extensions around the house.

- Check any equipment you intend to connect to your phone lines for its FCC registration number. This number is usually a decimal number followed by the letters *A* or *B*. You must call your local phone company business office and notify them that you intend to connect the item (phone machine, modem, etc.) and give them the registration number(s). Failure to do so can give the phone company the right (seldom exercised) to suspend or terminate your service.

- Use it before you buy it. Get a good working demonstration. With telephones, this means using it to call someone and listening to the quality of the reproduced voice with your ears. In other words, comparison-listen.

- Make sure the ringer equivalency number of the device you intend to connect is compatible with your phone system. If you attach too many or the wrong kind of phones or other devices to your phone wiring, none of them may work right.

- Check the warranty. Know where and how fast you can get the frebersatz repaired if you need to. Especially important for offbeat brands of cordless phones and answering machines.

In addition to the above checklist, answering machines should be checked for these features:

- Does it allow you to set the number of times the phone will ring before the unit answers?

- Can you vary the lengths of both the outgoing and incoming messages? Is the incoming message time voice-controlled (VOX)?

- Can you get your messages by dialing in from another phone? If so, does the phone machine let you know if there are no messages so you don't get billed for the retrieval call? Can you skip through messages when listening remotely? Will it give you only the newest messages on each remote call?

HINT: *Some* wireless telephones have been known to interfere with the disk drives of *some* common microcomputers. The reverse is also true: some micros can put out signals that interfere with *some* wireless telephones. There are hundreds of combinations, so the best thing to do is to test your new micro with your wireless phone in the location in which they will both be used. Don't overlook neighbors in adjoining apartments or homes during your on-site tests. Be considerate. If it turns out that these two machines won't get along in your softspace, there are many things you can do. Your local FCC field office can provide you with a booklet explaining how to avoid interference between devices. But the best thing is to test beforehand and just not buy equipment that interferes with other equipment.

_____4.4   THE PHONE LINES

The U.S. telephone system used to be a monolithic, coast-to-coast monopoly. During the 1950s and '60s, many battles were fought to break the monopoly Ma Bell had on phone manufacturing, phone service and installation. From 1975 on, under increasing pressure from consumer groups, potential competitors, and technological advances, the decrystallization of AT&T began to accelerate. By 1983 the nationwide system had been broken up into several regional companies, and a host of competitors had entered the telecommunications field. The resulting overall telephone system is a webwork of transcontinental telecommunications highways offering many options for getting from "here" to "there" in information space.

Most data communications can take place on _voice grade lines_ of the kind that you have been using for years with your prosaic vanilla telephone. Your telephone at home or office was designed to carry human speech, not microcomputer signals. Likewise, the lines that connect your softspace to a central switching office were meant for your voice. Hence, voice grade lines set limits on what can be accomplished in and through data communications networks. For most ordinary netweaving, these limits won't be significant.

Low speed does not necessarily mean poor performance. Data transmission speeds in the telephone network are technically limited to less than 9,600 baud [4.8.1]. This upper limit is seldom reached, though. Practical economic factors keep the upper speed at 1,200 baud for two-wire unconditioned phone lines—the normal phone service. Soon, 2,400-baud speeds will become more economical and therefore more widespread.

Figure 4.4 shows the general parts of the existing phone network, with phone company names on the various parts.

- The _local loop_ connects you, the user, with a _central office._
- The _interoffice trunk line_ connects the various central offices of a city.
- The _toll trunk line_ connects the system of central offices to a _toll office,_ which may be regional to several municipalities.
- The toll office connects to the thigh bone . . . no, that's not right . . . to the _high usage intertolls_ or _intertoll trunk_ lines. Note that it is this part of the overall system that is now being replicated by alternative phone companies: the long-distance services through microwave or satellite links.

4.4.1   Switched versus Private (Leased) Lines

A switched line is one the phone company rents to you on an as-needed basis. It has so many lines or channels going from city A to city B. You

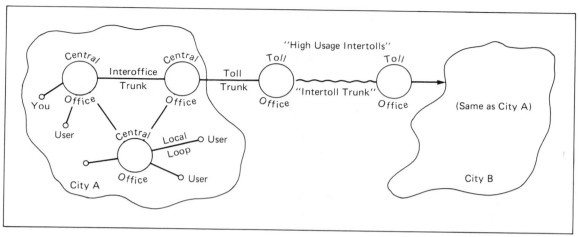

FIGURE 4.4. The Phone Line Network

in A and want to call B. When you dial up, you get whatever's available to take you there. If there is too much traffic, you may have to wait (this seldom happens, except on holidays or during local emergencies, when the trunk lines between cities or across country get jammed), and you may even get a busy signal. With private or leased lines this never happens. The line is yours, period.

4.4.2   Circuit Switching versus Packet Switching

Packet switching is a method of getting a digital message from point A to point B in a way that optimizes the use of both bandwidth [4.5.2] and individual channels. Think of the telecommunications channels as a railroad system connected at central nodes. In normal communications, each message is a single train. The "cars" are permanently linked to each other, and where the front goes, so goes the back. In circuit switching, all the connections necessary to get the message to its destination are made before the train ever goes on its way. Once the connections are made, the message is sent. All the connections must be maintained until both sides are finished with the channel. Then the connections are broken, and the channel is available for other messages.

With packet switching, the train is broken up into individual cars. The cars of one message can be interleaved with cars for other messages. New packets are added and sent all the time. This chain is composed of equal lengths of information, called packets. Each packet contains a specified number of characters and no more. In addition, an address section is put at the front of the packet. As each packet enters a switching point or node, the address information tells the node computer what the final destination of the packet is. Based on this address information, the node computer can send the packet on its way through the closest available channel. The advantage of this method is that it fully utilizes more of the available circuits more of the time, allowing a greater message traffic than would otherwise be the case.

The receiving computer, of course, puts the packets back together to form the original message with all the packets in their proper sequence. All this is done at speeds that essentially hide the process from the sender and receiver of the message. This is another way of saying that the system is "user transparent."

The two major commercial packet switching networks are General Telephone and Electronics' (GTE) Telenet, and Tymshare's TYMNET. These are the grandparents of the packet networks, because they were originally started in the 1950s to connect mainframes to one another and to remote terminals. They and networks like them may be called "data pack analog" networks because they convert digital information into analog signals [4.5] that are then sent on their way. In addition to these digital timesharing services, AT&T now provides what it calls Digital Dataphone Services (DDS) to do the same thing. DDS, however, use "true digital lines," which means that the transmission lines themselves are capable of handling digital signals, without the need to convert the signals to analog form before transmission. Eventually, we are told, all telephone lines will be truly digital transmission media.

Figure 4.4.2 shows a simplified packet switching scheme. The two messages are sliced into packets, here shown as single characters labeled with subscripts 1 and 2. The packets are interleaved and sent out in a continuous stream. Each packet is added to others and rerouted at each node of the network, depending on which lines have the least amount of traffic on them. This method allows the traffic on all the lines to be spread out, thus using all of them more of the time, which results in more economical use of existing facilities.

4.5 SIGNALS

We need to look more closely at the idea of a *signal* in order to understand how micros can be connected to one another and to larger computers by ordinary telephone lines.

Phone line signals are electrical. Your voice is an acoustic signal. Your vocal cords modulate the frequencies and shapes of vibrations of air. *Modulation* is using one signal or medium—in this case your vocal cords—to change another—the air. The concept of modulation of a signal occurs over and over again in telecommunications. Your voice vibrations, in turn, modulate electrical signals using the microphone in your telephone handset. The signals from the phone line are converted back into audible sounds at the other end by the miniature speaker in

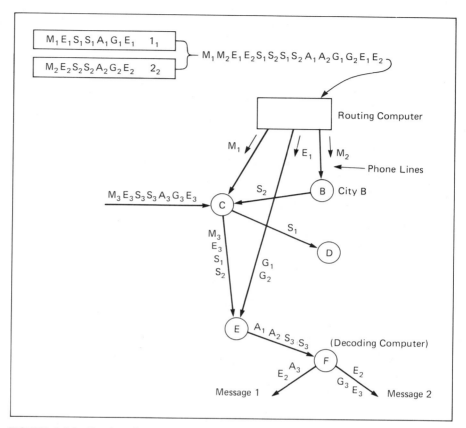

FIGURE 4.4.2. Packet Switching

the telephone earpiece. The air vibrations, in turn activate your eardrums, which modulate the hearing centers in your brain.

The signals used in transmitting sounds on the phone wires are called *analog* signals. If a graph of the variations in the frequency and volume of your voice is compared with a graph of the electrical signal, you will see *analogous* variations in the frequency (number of vibrations each second) and amplitude (strength or volume of the vibrations) of the electrical signal.

A different kind of electrical signal is used in microcomputers and similar devices. These *digital* signals consist of on-off pulses. No variations. Just the mere presence or absence of a voltage on a wire conveys information. These pulses, in a micro, can be used to represent a zero (0) or a one (1). This is handy, since micros do everything they do on the basis of zeros and ones, or *binary arithmetic.*

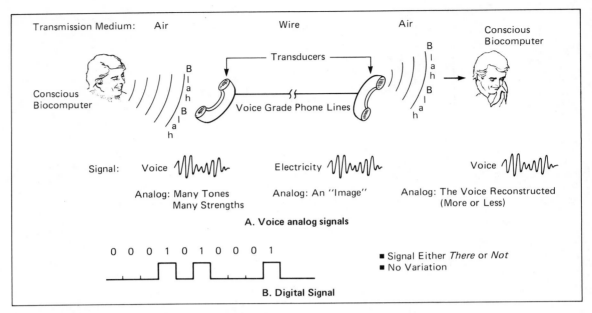

FIGURE 4.5. Electrical Signals

Figure 4.5b is a simplified picture of a digital signal as you might see it displayed on an oscilloscope.

4.5.1   Carrier Signal

The continuous audible tones exchanged by modems over voice grade phone lines are called carrier signals. If there is no carrier signal, no connection can be made. Likewise, if the carrier signal from either the transmitting or receiving modem disappears, the connection between the micros will be broken. Modem carrier signals are *sine waves,* fixed-frequency, unvarying signals shown in Figure 4.5.1.

The waves cycle from positive to negative and back again. The more cycles per unit of time, the higher the pitch (frequency) of the tone. By changing the carrier signal in predetermined and agreed-upon ways, it can be used to represent information. Digital information requires only two states, or two different kinds of alteration of the carrier signal, in its representation. Modulation accomplishes the change. Demodulation detects the change. A modem does both in a single device.

There are three different ways that a sine wave can be modulated to carry digital information: by altering its amplitude (the strength or voltage level of the signal), by changing its frequency (cycles per second or hertz, abbreviated Hz), or by fiddling with its phase angle (where the signal crosses the 0-point, relative to the previous cycle). Modulation of

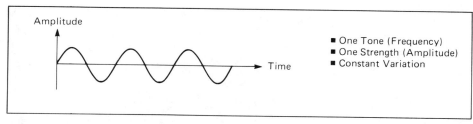

FIGURE 4.5.1. A Sine Wave

a signal is also sometimes called *keying,* which goes back to the telegraph and its clunky key. The two most common keying methods of interest to netweavers are frequency shift keying (FSK) and phase shift keying (PSK). Both are covered below in more detail [4.5.4, 4.5.5].

4.5.2   Bandwidth

Bandwidth is a measure of the size or capacity of a channel. This characteristic places limits on what can be transmitted through a given conduit. It is the difference between the highest and lowest frequencies that can flow through the channel. Figure 4.5.2 shows the relationships between the frequencies used in low-speed modems, the bandwidth of the telephone lines, the frequency range of the human voice, and the "buffer zones" or *guard bands* placed between the signaling frequencies. The

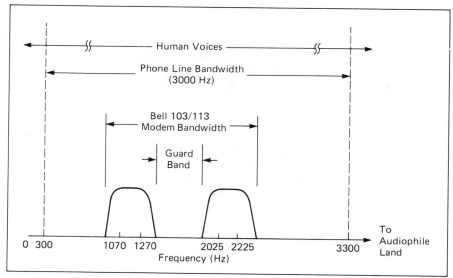

FIGURE 4.5.2. Frequency Bandwidths

greater the bandwidth of a given channel, the more information it can handle at the same time, and, usually the faster it can do so. (The human voice contains more information than is actually transmitted through the phone, as you can see.)

4.5.3   Digital to Analog to Digital

To recap, the sound you hear consists of a continuous range of frequencies, measured in cycles per second or hertz. Audible sound has a spread of frequencies—vibrations of the air against your eardrums—from about 25–30 Hz to 20,000 Hz. The human voice contains frequencies in this entire range. However, most of the audio signal generated by the human voice occurs in the range of 30 to 3,500 Hz. If you wanted to transmit hi-fi music over the phone lines, you would have to be able to send a continuous range of frequencies from (at least) 25 to 20,000 Hz. (Purists would claim you'd have to do better than that, since human beings can sense frequencies all the way up to 30,000 Hz.)

The human voice can still be understood (its information content has not been lost) if only the frequencies between 300 Hz and 3,000 Hz are transmitted and everything outside this range is ignored. This range just happens to fit in nicely in the bandwidth of the transmission channel that is your garden-variety voice grade telephone line. By limiting the bandwidth required to ship your voice-as-electrical-signal from place to place, the phone company is able to send many more conversations between central exchanges. This process is called *multiplexing* [4.8.4]. It results in great economies when sending signals over long distance.

In order to transmit your micro's internal digital signals over ordinary voice grade lines, you have to make the signal conform to the bandwidth requirements of the telephone system. So you have to have a device that creates an analog of the digital bit stream from your micro and sends it along the wires. At the other end, you have to have a similar device to take the analog signal and convert it to a digital form usable once again by a micro. The device that performs the conversion both ways is called a *modem,* for *mo*dulator/*dem*odulator [4.8].

4.5.4   Frequency Shift Keying (FSK)

FSK modulation uses two different frequencies to represent 0 and a 1. AT&T set up standards specifying which frequencies, at which speeds, would stand for which states in the answer and originate modes. The table in section [4.8.2] shows the four different frequencies used in FSK modems to represent 0s and 1s. FSK is used in 0–300 baud modems and is probably the most common method of digital communications today.

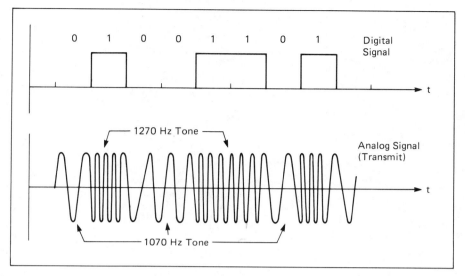

FIGURE 4.5.4. Frequency Shift Keying (FSK)

4.5.5   Phase Shift Keying (PSK)

PSK modulation is used in 1,200-baud modems. In Figure 4.5.5, the top graph shows two sine waves along the same time line. At any given time *t,* the signal is at a certain *phase* of its cycle. A complete cycle takes the signal through the values from 0 to 1 back to 0, to −1 and to 0 once again. This complete cycle can be divided into 360 parts or degrees. Think of it as a circle: the signal returns to 0 point every 180 degrees. If you add the negative part of the cycle, where the signal goes from 0 to −1 and back, then you have a complete cycle or one *rotation* of the phase angle through 360 degrees. The two sine waves shown are exactly 90 degrees *out of phase* with one another. The bottom graph of Figure 4.5.4 shows how a sine wave can be made to "jump" ahead or behind itself, creating a *phase change*. It is this change in phase that can be detected by appropriate circuitry and made to represent useful information.

4.6   MODES OF TRANSMISSION

You will find frequent references to *transmission modes, modes of transmission, communications modes,* and variations thereof. In data communications, several modes are used in transmitting signals from point to point. These descriptions of various modes may also apply to com-

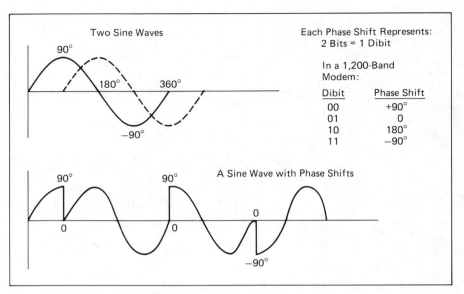

FIGURE 4.5.5. Phase Shift Keying (PSK)

munications between a micro and a peripheral device such as a printer or display terminal. The main modes of transmission are covered below.

4.6.1   Asynchronous

The main virtues of the asynchronous mode of data transmission are simplicity and low cost. Letters and numbers are defined by a code consisting of 5, 6, 7, or 8 bits, depending on which *character-oriented protocol* is being used [4.7.2]. Each character is "framed" by a start bit and a stop bit. Hence, you may sometimes hear this mode referred to as "start/stop communications." Each data bit is of equal duration, but the stop bit may sometimes be slightly longer than the others. Interpretation of the message depends greatly on the framing bits being received accurately. This mode is the one supported by popular micro modems, such as those from Hayes and Novation. The sending of bits does not need to be precisely timed. In this mode, characters can be sent as they are typed at a keyboard without having to be concerned about variations in typing speeds.

4.6.2   Synchronous

Synchronous communications depends on precisely timed bits and characters. Since precision timing and therefore more complicated and therefore more expensive electronics are involved in this mode of data communication, it is generally used only by large organizations on large computers. However, transmission speeds are usually higher with the synchronous mode. Data is broken up into groups or *blocks* of characters and sent as a continuous stream of bits. One of the modems generates a timing signal that gets both modems "in synch" with one another and establishes the basic data transmission rate. Special "synch characters" are sent in between each data block, to keep the modems dancing in step. Since data must be sent in blocks, characters typed from the keyboard are stored in a *buffer* (technically, a buffer is *any* data-holding area in memory) until the right number of characters has been collected and can be processed for transmission.

4.6.3   Binary Synchronous (Bisynch)

Bisynch is IBM's version of synchronous communication, in which the synchronization is achieved by signals generated at both the receiving and sending nodes. This mode is used primarily by IBM and its 3270 terminal clones.

4.7   SERIAL VERSUS PARALLEL TRANSMISSION

Figure 4.7 shows the basic difference between serial and parallel transmission. Inside your micro, a computer *byte* or *character* is 8 bits wide.* The internal communications channel (the "bus") of a micro uses signals that deal with all 8 bits at a time. This is called *parallel transmission.*

In our example, pretend that you are the letter *A,* a capital letter *A* in our standard English alphabet. (Are you boldface or italic? you ask. Please. One step at a time.) Before you can get zipped down the phone lines, you have to be changed into a form that the phone lines will handle. Pressing the keyboard letter *A* turns you into a binary number (equivalent to the decimal number 65). The agreement that specifies how you look as a binary number is the American Standard Code for Information Interchange (ASCII) [4.7.2.1].

Thus you have become a string of 0s and 1s, as shown in the diagram. Inside the micro, all your bits have been flashing about side by side, in parallel with each other, rather the way cars might travel down an eight-lane superhighway. Alas, the phone lines are a dirt track by comparison, capable of handling only one bit at a time, one after the other. This one-bit-at-a-time transmission is called *serial transmission.*

*I realize that a given micro may use a sixteen-bit processor and bus, or it may have a sixteen-bit processor and only an eight-bit bus (as in the IBM PC), or it may be one of the avant-garde thirty-two-bit machines. For purposes of explanation, I am assuming that one of the more common eight-bit micros will be what most netweavers end up using through the intermediate-term future.

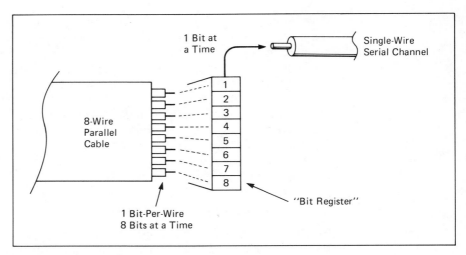

FIGURE 4.7. Parallel to Serial Transmission

The main advantages of parallel transmission are speed and simplicity. Serial transmission is not only slower but requires that we convert the parallel bits into serial bits before we can send them out. This means extra electronic circuitry. The chip that makes the change from serial to parallel and back again is called a UART.

4.7.1 Serial Cards, UARTs, USARTs, and SIOs

Printed circuit boards for converting the 8-bit parallel format into a one-bit-at-a-time serial format are called serial cards. At the heart of the serial card is a UART chip: Universal Asynchronous Receiver/Transmitter. Other electronic circuits designed to handle serial/parallel conversions are also used. The Universal Synchronous/Asynchronous Receiver-Transmitter (USART) and the Serial Input/Output circuit are essentially enhancements to the original design of the UART.

4.7.2 Character-Oriented Protocols

The protocols (agreements) that define how many bits are in a character and which bit sequence shall equal which characters are called character-oriented protocols. Character-oriented protocols are part of the *presentation level* in the OSI model defined by the International Standards Organization (ISO) [4.9.3]. There are several in national and international use. The most common are ASCII, EBCDIC, and BAUDOT.

4.7.2.1 ASCII
The American Standard Code for Information Interchange is by far the most universally used character set. It is even beginning to catch on

outside the United States. The ASCII set also defines special *control characters,* such as those for a carriage return or line feed on a printer. This standard was adopted in 1968 and uses seven bits to define 128 alphanumeric and control characters.

Some microcomputers (such as the IBM PC and PC-XT) use what is called an "extended ASCII" character set. This means that they add an extra bit to the code, making it an eight-bit protocol, and use the extra codes thus made available to define their own special characters, such as graphics elements.

4.7.2.2
EBCDIC

The Extended Binary Coded Decimal Interchange Code is used mostly by IBM on its mainframe machines. It has eight bits to a character. EBCDIC is used mostly in the binary synchronous (bisynch) mode of communication.

4.7.2.3
BAUDOT

The Baudot code has been in use since 1875. It is a five-bit code still used on teleprinters that are widely used in the deaf community. However, the use of ASCII is gradually displacing baudot.

4.8   MODEMS

The modem (modulator/demodulator, also called a "signal converter") is the crucial link between your personal information system and the global information environment. There are many kinds of signal converters. The ones netweavers are concerned with link micros to the phone lines.

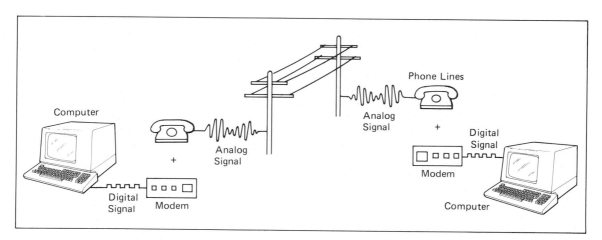

FIGURE 4.8. The Modem Connection

4.8.1   Baud Rate

The basic specification of a modem is its baud rate: how fast the signals can be sent back and forth. The phone lines themselves impose limits on this rate, but these limits are not significant for most of the work a netweaver will be concerned with. Technically, the baud rate is defined as *the number of discrete signal changes per time unit* that can be sent/ received on a given channel. Naturally, this rate will vary according to the bandwidth of the channel, the method of encoding (modulation) used, and so on.

The term "baud" comes from "baudot," after an early pioneer in telegraphic telecommunications, Emile Baudot (who also gave his name to the Baudot code). When the term first came into use, it served solely as a measurement of how fast dots and dashes could be sent on telegraph wires. Dots and dashes only require that two states or changes be detectable on the channel. As with many terms now used in the information field, the definition of "baud" has shifted slightly with the shift in the underlying technology from the electromechanical telegraph to the all-electronic computer.

In low-speed modems most commonly used, a modem's baud rate will be the same as the number of bits (on and off pulses) that can be sent each second. The changes being detected are changes in tone or frequency (frequency modulation, FM or FSK). Thus, *in this case,* a 300-baud modem speed translates to 300 bits per second for the micro to use, which is roughly equivalent to 30 characters per second, or about 350 words per minute. If you're an average reader, this is a comfortable speed for your biocomputer to process information coming in on a video display terminal. At 300 baud, words appear on the screen about as fast as you can read them.

However, to shuttle large amounts of information (say a chapter of a book or an entire book to a typesetter) between your softspace and a mainframe somewhere across the country, you'll need to increase the exchange speeds. The current upper limit on voice grade lines is 9,600 baud. However, there is an economic limit that keeps netweaving activities in the 1,200-baud and lower range. This is because as the baud rate increases, so does the complexity of the circuitry required, and hence its price. High-speed modems gain their speed by using a more sophisticated method of sending bits in groups of two (dibits), three (tribits), four (quadbits), or more. If each signal change represents only one bit, then you need only two different kinds of signal to send strings of zeros and ones. However, if each signal change is going to represent two bits, then you need at least four discrete kinds of detectable signal to determine which group of two (00, 01, 10, or 11) is being sent.

The two most common baud rates are 300 and 1,200. There are still some very slow modems using 110 baud out there, and you can still buy these, too. Specifications for modems were set a long time ago by the Telco (Ma Bell, a.k.a. AT&T, a.k.a. AIS). The tables given here lay out the basic differences between the various specifications. For each type of modem, a particular set of tones (frequency, specified in Hz) will represent the value zero (0). A second set of frequencies will represent the value one (1).

- A Bell 202 [CCITT V.23] modem can communicate *only* in 1,200-baud half-duplex mode.
- A Bell 212A [CCITT V.22] modem can communicate at 300 *or* 1,200 baud, full duplex.
- The 212A modem is the preferred standard.
- The Bell 113 modem is originate-only. It is not generally used in personal softspaces with microcomputers.

4.8.2   Answer/Originate

Notice that in the set of tables giving the Bell specifications for modems there are two sets of frequencies labeled "Answer" and "Originate." Each modem has to be able to distinguish its own micro's bits from the bits sent by the machine on the other end. By convention, when you call up a data base or conferencing system or whatever, the machine that answers your incoming call has already set its modem to the "answer" mode and is waiting for a call. That means that yours will automatically have to be the "originate" modem. Technically, it doesn't matter which set of frequencies either modem uses, so long as they are not the same.

When you connect your micro directly to another over the phone lines, you have to decide in advance which of you will be "answer" and

Frequency Shift Keying

Bell 103 [CCITT V.21]		Answer	Originate
Transmit	1	2,025 Hz	1,070 Hz
	0	1,225 Hz	1,270 Hz
Receive	1	1,070 Hz	2,025 Hz
	0	1,270 Hz	2,225 Hz

Phase Shift Keying

Bell 212A [CCITT V.22]	Answer	Originate
Transmit	2,400 Hz	1,200 Hz
Receive	1,200 Hz	2,400 Hz

which "originate." The simplest thing to do is have the person who dials the number specify his micro as the "originator." The other one can answer.

The same applies if you are connecting two machines together in the same location using a null modem cable [4.13.2] or two modems directly connected together. You must specify in advance which micro will be the "originator" and which the "answerer." If both are set the same, no communication and a lot of frustration will result.

4.8.3 The Modem-to-Phone-Line Connection

There are two basic ways for a modem to connect with the phone lines: acoustically, using sound as the coupling medium, or electrically, using wires or cables as the only coupling medium.

4.8.3.1 ACOUSTIC-COUPLED MODEMS

Acoustic-coupled modems use a standard telephone handset (as opposed to, say, a more stylish but less functional handset, as on Princess and Slimline phones or phones shaped like ducks) to send and receive signals. It has two foam cups into which the earpiece and mouthpiece of the handset are inserted. This type of modem uses a small speaker and microphone of its own to "acoustically couple" the phone to the modem. The purpose of the foam cups is to channel the sound and to block stray noises from entering the signal. This kind of modem is generally low-cost (its main advantage) and was in widespread use in the early 1980s. The most frequent error encountered when using acoustic-

FIGURE 4.8.3.1. Direct-Connect Modem
Photo: Courtesy of Hayes Microcomputer Products, Inc.

coupled modems is the inadvertent misplacement of the telephone handset: putting the microphone of the modem next to the mouthpiece of the handset, and thus putting the speaker part of the modem next to the speaker part of the telephone. This error can be very disconcerting until it is discovered. Aside from cost, another advantage of this design is that you can use it in places that may not allow you to direct-connect a modem to the phone lines. Generally, acoustic-coupled modems are fine for teleconferencing, BBSing, or database access, but avoid these modems if you expect to exchange programs with another system and expect them to run flawlessly once exchanged.

**4.8.3.2
DIRECT-
CONNECT
MODEMS**

Direct-connect modems overcome the limitations of acoustic-coupled designs. They plug directly into the phone lines through the wall jack or the telephone set itself. The signals exchanged are thus much cleaner, since all extraneous environmental sounds are eliminated. Direct-connect modems can usually respond to incoming ringing signals, making them "auto-answer" as well as direct-connect. They may also be able to autodial, meaning that the modem itself can generate dialing signals that the phone company switching centers can recognize and respond to. With the proper software, direct-connect, auto-answer, autodial modems can help automate your telecommunications activities.

4.8.4   Multiplexing and Concentrators

Multiplexing is a method of making a given channel (which would normally serve only one sender and one receiver) serve many users at the same time. Multiplexing equipment is sometimes referred to as MUX

FIGURE 4.8.3.2. Acoustic-Coupled Modem
Photo: Courtesy of Novation, Inc.

hardware or just MUX. A concentrator is essentially the same thing: taking many signal sources and concentrating them over a single channel. There are plenty of ways to do this, all of them expensive.

4.9   PROTOCOLS AND STANDARDS

Protocols are rules or conventions set up for purposes of data communications between a micro and its attached appliances, or between two micros. A "standard" is a protocol that is *supposed* to be universally accepted. In that sense, however, there are few actual standards in the telecommunications industry. Some of them approach *nearly* universal acceptance. The emphasis is on *nearly* because loud claims are made for standards that are nowhere near being universally applied. For example, promoters of IBM PC micros will claim that IBM "set a standard." Actually, it simply created, out of whole cloth, a standard based on its ability to get a lot of people to buy its machines. Needless to say, theirs is not the only microcomputer around.

You will also see references to "de facto standards," which means that enough people use them to make them "standards," at least for that group. The problem is that you may not want to be a part of that group, for various technical and political reasons. In that way, then, the lack of standards is an evolutionarily positive thing, for it means that there is still room for invention. Standards set too early can become straitjackets to further progress. Witness the sad state of television image production. It is possible to have a much better television image than we actually have in the United States, but our technical "standards" make upgrading the image a very expensive—hence opposed—proposition. We have the technical ability but lack the political-economic will.

As another example, there has been much touting of Unix, AT&T's internally developed operating system, as a de facto standard. It is not, in fact, a standard at all. What is more, a "standard" operating system of any kind is unlikely in the near future, for a variety of reasons. Any good newcomer in the field of operating systems generally is thus in the running for the title of "standard," and has the potential of becoming widely used and accepted. Examples like this can be multiplied throughout the telecommunications industry.

4.9.1   Handshaking

The exchange of prearranged introductory signals between two devices is sometimes called handshaking. Thus you may run across references to "handshaking protocols." Handshaking is another one of those shaky terms: it can refer to:

- Hardware handshaking: control signals and acknowledgments controlled by hardware
- Software handshaking: signals controlled by software
- Log-on handshaking: referring to passwords and terminal codes required from a user to allow the host computer to determine who and what kind of terminal has logged on

Protocols specify in great detail how a message is to be packaged and sent. There are protocols such as ASCII that specify which sets of bits will stand for which characters. There may also be protocols to define a special meaning for sequences of characters. For example, the sequence ACK sent after a message has been exchanged may ACKnowledge to the transmitter that the message got through. Many entities have had a hand in setting up protocols: the phone companies of course, hardware and software manufacturers, international and national regulatory or standards organizations, and others. Generally, all components within a single system must use the same protocols to achieve communication. Alternatively, some method of converting from one protocol to another must be provided in cases where the communicating devices use different conventions.

Protocols also specify such pertinent items as whether or not the transmission is to be full or half duplex, the character set to be used (ASCII, EBCDIC, BAUDOT), and which frequencies will be used to represent 0s and 1s.

4.9.2 Break

Many remote terminals and micros have a key labeled BREAK or BRK. This key does not represent any of the standard ASCII characters. Rather, it generates its own special signal for transmission to the outside computer system. This allows the signal to be sent at any time, even during the transmission or reception of blocks of data. What the other system then does will vary from computer to computer. Sometimes this break protocol is not implemented, and nothing will happen at all when the break key is pressed. This is rare, though.

The break signal for each given computer system will be defined as the *minimum* amount of time that the break key must be pressed and held before an action takes place. In its most common use, pressing the break key means that you want to take control back from the other computer. For example, let's assume that you asked for a given file to be transmitted to you for printing on your screen or on your printer. When the other system complies and begins its "dump," you discover that it wasn't the file you wanted; or you discover that it's much too long for your buffer to handle, not to mention the cost of staying on-line for that length of time. Pressing the break key will interrupt the transmission and allow you to issue a new command.

4.9.3   ISO Network Reference Model

This is a guide, established by the International Standards Organization, for the configuration of networks of all kinds, including local area networks (LANs). It is called a *layered model* because the system is built from low (physical) to high levels in well-defined shells, rather like the layers of an onion. The first, or lowest, two layers are hardware specifications. These two layers determine such things as the topology of the network and its components. They also determine the bandwidth and the access mechanism to be used.

The bandwidth specification is one that has direct bearing on the speed (in LANs, this is usually expressed in millions of bits, or megabits, per second: Mb/s) of the network, which, in turn, will establish limits on how many stations, users, or devices can use the network at the same time. Speed or bandwidth alone is not an accurate indicator of overall network performance, but it is the specification that closely defines its overall power, especially for LANs. Bandwidth figures indicate the amount of data that can be transmitted through a given channel over a given unit of time.

The time it takes to send a message between two devices (micros, mainframes, nodes) is called *effective throughput,* and this speed is affected by both hardware and software factors, not just the bandwidth of the network. The effective throughput rate of networks is usually slowed to the speed of the micro(s) involved, which can be considerably slower than the bandwidth will allow. As we have seen, the effective throughput of networks using voice grade lines is limited by the bandwidth of the telephone network.

As another example, a LAN operating at 1Mb/s will take less than half a second to load a 40K applications program from the hard disk to the local user station. A .5Mb/s network would also take less than a second to load the 40K program, but it would be significantly cheaper to implement than the 1Mb/s net. Finally, a transfer of 5 megabytes of information from one hard disk drive, through a micro, to another hard disk drive at a rate of 9,600 baud will take around five and a half hours to accomplish.

Two major micro architectures prevail at the lowest (physical) level, which is where LANs are connected: Ethernet, established by Xerox Corporation, and ARCnet, a product of Datapoint. The ability to link LANs together in multi-LANs is an important long-term goal in the industry, so going with one of these currently popular LANs will probably assure long-term interconnectibility. However, since no universal agreements concerning networking standards have been adopted, and since the technology is still in a state of flux, something better may come down the pike soon. For more on LANs, see [4.16].

The ISO model is shown in Table 4.9.3. The first four levels are more widely followed than the last three. There are four competing protocols at layers 3 and 4: XNS [4.17.1.2], SNA [4.17.1.3], IP/TCP [4.17.1.1], and ECMA [4.17.1.4].

4.9.4   Networking Standards

Netweavers should be particularly wary of any "protocol" or "standard" that is not in the public domain, that is, one that's proprietary to a particular company or group of companies. Many software producers have come up with communications protocols suitable for use on networks—either LANs or the phone networks. There are two reasons why a company goes with a private, self-produced protocol, neither of them viable

TABLE 4.9.3   The OSI (Open Systems Interconnection) Model

Layer 1—Physical Level. Defines the interface between a terminal and its data communications equipment. At this level, mechanical and electrical definitions are set forth. Cables, connectors, voltages, and modulation techniques make up the protocols for this layer. This is the level at which the RS-232C standard is defined.

Layer 2—Data Link Control Level. Specifies what the "message unit" will be, its format, and the means of network access. On LANs, message units are usually called *packets,* and each packet is given an address code (to which device or node the packet is being sent) and an error-detection checksum. Binary synchronous communications protocols define this layer.

Layer 3—Network Level. Defines the actual route of data across the network. It also defines network management functions, such as the delivery of status messages to nodes and the regulation of packet flow.

Layer 4—Transport Level. Sets the distribution of addresses of the various message recipients (nodes), error detection, and recovery.

Layer 5—Session Level. Defines the making and breaking of connections and the actual exchange of data.

Layer 6—Presentation Level. Defines the syntax, transformation, and formatting of data. The North American Presentation Level Protocol (NAPLP) for videotext graphic representation is defined at this level. The syntax and format of an application will usually differ from the syntax and format (packets) used in transmission, and this layer provides the translation between the two.

Layer 7—Application Level. Defines the mechanisms and methods by which applications enter the OSI model. It supports user applications but is not itself any of those applications. This level includes log on utilities, password protection, file transfer and the like.

from the netweaver's point of view. The first is to require all the machines in a given network to use the same software, the same family of software, and, usually, the same vendor or consortium of vendors. The rationale, from the company's point of view, is that there will then be more sales of a given communications program or family of programs. The other reason is to be able to license other producers who wish to make devices or software that will interface with the network using a particular protocol set. This is an additional source of revenue for the company. The cost of such licenses usually precludes individuals or small companies wishing to develop their own networking software primarily for their own use.

Not only should protocols be machine-independent, they should also be manufacturer- or supplier-independent.

The eventual network standards to be adopted should have the following characteristics.

- They will be machine-independent.
- They will be manufacturer-independent.
- They will be in the public domain.
- They will follow the OSI model as closely as possible.
- They will offer data transparency.
- They will provide flow control.
- They will provide error detection and recovery using a CRC (Cyclic Redundancy Check) algorithm as derived from CCITT specifications.
- They will provide for packet transmission.
- They will provide a virtual file format so that dissimilar machines can transfer files between them easily.
- They will provide for data encryption that is more secure than the DES (Data Encryption Standard: [4.10.4]).
- They will be efficient on even small machines or terminals.
- They will use the NBS (National Bureau of Standards) protocol data units scheme.

4.9.5 NAPLP

The North American Presentation Level Protocol was developed in Canada on the Telidon system. It provides for a method of transmitting graphics on the phone lines easily and with great results. Why was that a problem? you might ask. Because to encode visual information other than letters and numbers requires much more in the way of bandwidth to transmit. An ordinary television picture, for example, contains orders of magnitude of more information than the same screen full of text. Since the bandwidth of the telephone lines is limited, there needed to be some way of sending the amount of information required by even still graphic frames. You could send it, of course, using ordinary means, such as slow-scan television signals. But this takes a long time. NAPLP provides a means of creating and sending graphics quickly. AT&T is supporting this standard, and a number of micro manufacturers, peripheral manufacturers, and software companies are developing NAPLP capabilities for various micros. NAPLP is not the only graphics standard being promulgated. A second, developed in Europe, is being supported by IBM. However, NAPLP is generally a better-quality system for getting images into your networks.

4.10 ERROR DETECTION AND RECOVERY

In highly sophisticated computer systems that control real-time events (the Mission Control computers at NASA, for example), it's essential that what is received is what was sent and, furthermore, that what was sent is what was *meant*. In life-support systems, a wrong bit at the wrong time can mean death, so equally sophisticated methods have been developed for assuring that wrong bits are never generated, and, if generated, are never sent, and, if sent, can be filtered out before being acted upon by the receiving system. Even so, accidents happen. There is no such thing as "fail-safe" in systems designed by human beings. We are not sure that there are fail-safe systems *anywhere* to be found, even in nature.

For most common purposes, though, where most netweavers will be operating most of the time, 99 percent accuracy is adequate. However, if you're dealing with financial data in large volumes, or transmitting price lists, stock quotes or any other information that people on the other end are going to be basing crucial decisions on, then you are going to want 100 percent accuracy. In data communications and netweaving work, there are ways to approach this degree of reliability.

Table 4.10 shows the types of error detection and the degrees of

TABLE 4.10   Error Checking and Degrees of Accuracy

Degree of Accuracy	Type of Error Detection
Low: 85% (very noisy channel) ↓	No parity / Simplex No parity / Full Duplex Parity Checksum (Block) Parity + Checksum + Echoplex
99.9% 99.999–100%	CRC (parity not used) Hamming Codes

accuracy associated with each. The low of 85 percent is associated with forms of transmission in which virtually nothing is done either in software or in terms of line conditioning to maintain accurate transmission of bits. Thus, if the line is particularly noisy, or if the carrier signal strength is reduced, 15 percent of your text may show up as garbage. If the text is a letter or memo, or other textual information you're downloading for one-time use, there may not be a problem, even at 85 percent accurate. Why? Because you'll probably be able to read and make sense of the text even with the errors in it.

4.10.1   Redundancy

In most forms of communication more is sent than is strictly necessary to convey the desired information. This extra information makes it easier to make sense of the message. In information theorists' terms, the message has become more predictable or *redundant*. Messages written in English text—or any other human language for that matter—are never completely unpredictable, never completely a surprise to the receiver. If they were, they wouldn't be messages at all, but just noise. The rules of the language (its protocols: spelling, syntax, structure, and so on) are information shared between and known in advance by sender and receiver.

Nearly every passage of prose, from the Bible to the gaudy patter of detective novels, can be edited and still give the reader the essential message of the author.

Nearly every passage of prose can be edited and still give the reader the essential message of the author.

Nearly all prose can be edited yet give the reader the author's message.

Nearly all prose can be edited. And sometimes the author's message is lost.

A single typographlical error does not—cannot—substantially garble the meaning of this sentence. We tend to tolerate a certain amount of typographical glitch in most printed material because we are able to "fill in" the missing or erroneous pieces owing to the amount of redundancy in our language. Some studies have shown that this "tolerance" level is higher with video display screens than with paper.

4.10.2   Parity

To review: This is a bit, 0 and this is another bit, 1. Using bits, and bits alone, we can create codes that can handle just about any kind of information in existence.

In the world of micros, *two* bits is called a *slice:*

00

01

10

11

As you can see, there are only four possible slices allowed. However, if we take the slices two at a time, we create what is charmingly called a *nybble* (I'm not making this up, you know!). There are sixteen possible nybbles in the micro universe.

Taking the nybbles two at a time, we get the much-touted *byte.* The byte, which is eight bits long, gets all the attention, partially because it has become a de facto unit of measurement for various kinds of memory. It also happens that much of the internal work done by micros is done eight bits at a time, over *parallel* eight-bit channels. Sometimes this eight-bits-at-a-time communications is called *bit-parallel* or *byte-serial* (bite cereal?).

A minimum of seven bits is required to translate text information into digital data. Why? Because there are twenty-six letters in the English alphabet. There are ten digits, and a variety of special characters such as the comma or the hyphen or the plus sign. If you add lowercase letters, that's another twenty-six. Okay. If we tried to encode this set of required characters using bits or nybbles, we'd run out of possibilities before we were able to cover all the characters.

The ASCII [4.7.2.1] code is a seven-bit character-oriented protocol. It gives us 128 characters. Why seven and not a byte (eight)? Because the American National Standards Institute (ANSI) says that seven is the minimum number needed to support most applications, that's why (ANSI X3.4-1977 standard). Stop asking impertinent questions!

Yes, but what's *parity*? I'm getting to that. Really.

Here's the letter *A* as it appears in ASCII:

A = 1 0 0 0 0 0 1

These are called the *data bits.* They tell us what we need to know. If we send them, one at a time, serial fashion, through the phone wires, the computer at the other end can work its interior voodoo and display the letter on the receiving screen.

But what if one of the bits "gets lost" on the way? Or gets changed from a 0 to a 1 or vice versa? How will we know? Our *A* can get transformed into some other letter and we wouldn't know it. Here's where redundancy comes in. What we do is *add some more information* to the data bits that will serve to correct changed bits. This extra information can be added using only one more bit, called a *parity bit.*

Parity is *information about information,* in this case information about the outcome of a procedure performed on the data bits. The procedure is called *checksumming.* The parity bit is calculated by adding up—checksumming—the data bits. Remember that in the English language redundancy is built in by virtue of our shared information about grammar, syntax, and spelling. So, too, parity information is set in advance, by agreement between the communicating parties. The agreement specifies whether parity is to be *odd* or *even.* Table 4.10.2 should help make this easier to grasp.

The checksum is performed on all eight bits, including the parity bit. If the pre-agreement is that parity will be odd, then any checksum that comes up even on the receiving end will signify that there has been an error in transmission. The receiver can then request that the sender retransmit the offending character.

Parity is sometimes also specified as *low* (odd) or *high* (even). Sometimes there is no parity at all.

Parity as a form of error checking is one of the simplest forms of detection and correction schemes. It relies on the statistical fact that having two bit errors in a given transmitted character is a rare occurrence. If *two* bit errors per character should happen to occur, though, parity checking won't detect it.

4.10.3   Other Redundancy Error Checking Techniques

In asynchronous communication [4.6.1], parity checking is the most commonly used form of error detection and correction. Other techniques may be used, especially in synchronous [4.6.2] communication. These include *longitudinal redundancy checking* (LRC), *vertical redundancy checking* (VRC), and *cyclic redundancy checking* (CRC). Error checking can be performed by either hardware or software. Error checking protocols, whatever they might be, are easily built into sophisticated communications software packages.

4.10.4   Public Key Cryptosystems and DES

Error checking techniques and the principles of redundancy are part of maintaining *data integrity,* which is part of the overall security of an information system. Closely allied with data integrity is the problem of

TABLE 4.10.2   ASCII *A* with Parity Bits Added

Data Bits	Data Bits + Parity Bit (Odd)	Data Bits + Parity Bit (Even)
1000001	10000011	10000010

data security. The object of the game is to make sure that unauthorized people cannot get to your information if you don't want them to. Also, when transmitting information on the phone lines, we need to make sure that sensitive information is not intercepted, interpreted, and used by those other than intended by the transmitter. This is especially important when we begin to talk about the mail metaphor: if e-mail is *really* mail, then it has to have some of the characteristics that we associate with the transmission of a sealed, first-class envelope containing a letter to a loved one or business associate: a means of enclosing it so that no one else can read it and a means of signing it so that the receiver *knows* whom the letter came from.

The Data Encryption Standard (DES) published by the National Bureau of Standards (NBS) in 1975 and developed by IBM is an attempt to meet the first need: the provision of a "privacy envelope" for digital messages. Data encryption uses the same methods employed by governments and their intelligence agencies to make sure secrets remain secrets.

Cryptography protects information by changing the original information, called *plaintext,* into coded or encyphered information, called *cyphertext.*

The DES has been widely criticized by cryptographers as being too easily broken by large machines—the kinds of computers you can't lift, the ones owned by governments and large companies. Controversy in the academic community over the security, in the long run, of DES makes it wise for netweavers to demand something better. The more paranoid among them have even suggested that IBM, in collusion with the National Security Agency (NSA), may have built a "trapdoor" into the DES standard that would allow anyone who knew about it to easily break the cyphertext generated by DES. Who knows? In the short run, however, DES is better than nothing, because after nearly ten years of scrutiny by experts, no cheap, viable method of cryptanalyzing the DES has been discovered or published.

The *key* to a cypher is a number or set of numbers that lock the plaintext and unlock the cyphertext. DES uses a key that is 56 bits in length. There is only one key, and that key must be kept secret, shared only by the transmitter and the receiver of a message. If the key is compromised in any way, it has to be changed.

Part of a solution to the problem of the security of encyphering keys is offered in so-called public key cryptosystems. Each of us would publish, in a directory, a *public key* that is used to encode information by our correspondents. Once encrypted, the message cannot be decoded using the same public key. It can only be decoded using a *private key,* which is not published. Thus, the key that locks the message can be used by anyone, but the key that unlocks it is only in our possession.

Until we can ensure secure communications, including methods of encryption and verification that are low-cost and easily implemented in mass-produced devices such as microcomputers, there are no technical barriers at all to invasions and erosions of privacy. Attacking computer crime with legislation alone fails to address this fundamental problem.

_4.11 COMMUNICATIONS CHANNEL CHARACTERISTICS

Communications channels can be classified yet another way: according to the direction(s) and ways in which they are able to move information. A given channel may be *simplex, half duplex,* or *full duplex.*

4.11.1 Simplex

A simplex channel allows transmission in one direction only. Most mass media use simplex channels: broadcast television, cable, radio.

4.11.2 Half Duplex

A half-duplex channel allows two-way communication over a single channel. Each user must "give" the channel to the other in order to change from transmit to receive. Citizen band (CB) radio is the most common example of half-duplex communication. In half-duplex data communication systems, the micro or terminal must wait for the other machine to stop transmitting before it can send anything in turn. Data cannot be transmitted and received simultaneously.

4.11.3 Full Duplex

Full-duplex channels allow simultaneous two-way conversations. An ordinary telephone conversation between two people is an example of full-duplex communication.

4.11.4 Echoplex

Echoplex is a feature of full-duplex communication and refers to how the data are treated by the two devices involved. A transmitted character is "echoed" by the receiving computer or terminal, and the character is then displayed on the transmitter's screen. If a character is lost, it won't show up on the transmitting screen at all. If it is distorted, it will show up as a different character on the screen. Echoplex is thus a method of error detection albeit a simple one.

Echoplex is usually used in all full-duplex data communications, so it is common to use the terms "echoplex" and "full-duplex" interchangeably.

_____4.12   MICROS, TERMINALS, AND MICROS AS TERMINALS

In the early days of microcomputers, it was valid to ask, "What do you want to do with it?" Now a valid answer is, "Everything." One thing, though, is becoming a basic feature of all micros, no matter what the size or price range: telecommunications. Thus all of the current and future micros either have modems built in or can be easily outfitted with one.

At the lowest end (in terms of information-handling capability and usually, but not necessarily, in terms of cost as well) are hand-held terminals, inexpensive (under $200) microcomputers, and specialized black boxes of the videotext and teletext variety. The only reason to buy one of these devices is size. If you do a lot of traveling, you'll want something lightweight and inexpensive. Most of them allow little other than accessing the central data bank. There is not much one can do with these in the way of preparing text and/or graphics of any but the most rudimentary kind. They do provide easy access to the existing network nation, however, and may be just fine for certain kinds of training application.

In the intermediate range are so-called dedicated terminals and the various home/business micros outfitted with appropriate modem and software that makes them look like dedicated terminals. Here there is a lot of variety. The systems approach to choosing and using these intermediate-range systems is recommended [1.6, 3.3.2]. Terminals are not as expensive as micro systems, but they are not as powerful either. More costly than black boxes, terminals are usually used in local situations where many users must be connected to a system in the same building. They are not generally recommended for remote access to networks or information services. But as with all rules of thumb, there are times when a terminal is called for, either for economic or logistic reasons.

At the upper end of capability and price are micro systems that let you do a wide variety of local processing, information creation (graphics frames and text and mixtures of the two), and transmission using the same system. Generally, this means a microcomputer that has at least 256K of RAM and 1Mb of storage. In addition, you'll want a bit pad for graphics, a card or cards for creating and accessing encoded graphics information (NAPLP or better graphics protocols), and at least 1,200-baud modem capability.

_____4.13   CABLES

Someday all our mechanical and electronic devices will talk to each other without using bundles of wires strung all over the place. Until then, netweaver, you'll have to face, sooner or later, the bugbear of all systems designers: cables and the lack thereof. I have seen grown people shed tears and burn incense to long-forgotten gods and even then come close to adding straitjackets to their permanent wardrobes, all for the lack of handshaking between a printer and a microcomputer connected by what appeared to be an ordinary, innocent, run-of-the-mill "standard" RS-232C cable and its attendant connectors.

Somehow, though, it makes perfect system*antics* sense. You put together hundreds or thousands of dollars' worth of micro equipment, including printer, monitor, keyboard, modem, graphics pad, music keyboard, math coprocessors, and whatever else. Then you can't get the pile of parts to work properly because of one of the least expensive components of the system: a $20 cable. Sigh.

As a netweaver—even as an ordinary wary *consumer*—you would be well advised to arm yourself in advance against the wiles of cables, and *especially* the RS-232C "standard." In the words of one accomplished netweaver, "All RS232 really means is that you can usually cross-connect one cable wire to another without seeing a puff of smoke, and that signals are not guaranteed over transmission distances greater than fifty feet!"

At minimum, never assume that two devices will be able to talk to one another, even though the salesperson or the advertisement for the product(s) said they would.

Find out if the cables are included in the price of purchase for a bundled system (products sold together as a system and meant to work together), or if they're extra.

4.13.1   RS-232C

This arcane designation is used in a lot of advertisements in magazines and in manufacturers' literature. The RS-232C standard was set by the Electronics Industries Association (EIA). It is part of the physical level protocols [4.9.3] for networks and/or communicating devices. It specifies in detail the physical characteristics of the connectors to be used as well as the signals that must be present on each pin of the plugs. However, most manufacturers do not implement the full set of functions on their "RS-232-compatible" systems. This is why it is important to check out the actual specifications as supplied by the manufacturers before assuming that any given device will communicate with another using an RS-232C connection.

4.13.2   Null Modem and Crossover Cables

A null modem cable, also called a crossover cable, is an RS-232C cable wired in such a way as to "trick" the two micros it connects into talking with each other. It "crosses" two or more of the signal lines so that no modem is needed between the two micros. The wiring diagram for this kind of cable is given in Appendix IV.

4.13.3   Centronics

The Centronics standard was developed by the Centronics company, a manufacturer of printers. This standard is for parallel digital communications. Even though it is not used in phone line data communications, it is included here because no discussion of cables would be complete without mentioning it. Sooner or later, a netweaver will come across this standard in his or her work. The Centronics physical level protocol became a de facto standard because a large number of their printers have been installed in the field. In general, a cable that is wired according to the Centronics standard will work every time with devices that are advertised as "Centronics-compatible."

4.13.4   RS-449

The EIA issued RS-449 ("General Purpose 37-Position and 9-Position Interface for Data Terminal Equipment and Data Circuit-Terminating Equipment Employing Serial Binary Interchange") in 1977, a scant four years after the RS-232C standard was adopted. RS-449 was devised to overcome the limitations of—yet remain somewhat compatible with—the RS-232C standard. Related updated standards include RS-422-A: "Electrical Characteristics of Balanced Voltage Digital Interface Circuits"; and RS423-A: "Electrical Characteristics of Unbalanced Voltage Digital Interface Circuits." Perhaps someday all new equipment will follow these standards. For now, it's touch and go.

Even though the RS-449 specification is technologically *better* than the older RS-232C, there is a built-in, systemic resistance to change, simply because there is so much RS-232C equipment in place that is still usable. As this is being written, we are waiting for an inexpensive "black box" adapter that will go between the two.

But *why* is RS-449 "better"?

- Because it can handle higher speeds.
- Because transmission distances can be greater than fifty feet.
- Because it, unlike its RS-232 predecessor, adequately specifies a connector for the standard.
- Because it's better adapted to integrated circuit design.
- Finally, because it improved the electrical characteristics of the standard by providing for *balanced circuits*. (For more precise details on the advantages of balanced versus unbalanced electrical circuits, see the reference in Appendix IV: [4.5].)

4.13.5   IEEE-488

This standard is another parallel interface specification. It is used in Commodore computer equipment. The only way to use non-Commodore modems with machines such as the VIC-20 or the Commodore 64 is to get Commodore's conversion kit, which converts the parallel output to a serial form that other modems can use.

4.14   SOFTWARE: PROGRAMMING AND NETWARE

To adequately coordinate and manipulate information within your soft-space(s), you'll need more than the phone lines, micro system, modem and cables; you'll also need software of various kinds. There are three ways to go about making all the components of your telecommunications system work together:

- Programming the system yourself;
- Getting some other programmer to customize your system;
- Buying off-the-shelf packages as you need them.

The more you are able to do yourself, the better. But there is no shame in buying off the shelf or in hiring someone else. Just expense and greater expense. Prepackaged software is coming down in price and going up in performance just the way hardware has been doing for these past few years. That trend certainly will continue.

The software concerned with modem control, information retrieval from remote data bases, protocol conversions, and the like has been called telecommunications or just communications software. Taken as its own category, there are three subdivisions possible in classifying communications software according to this traditional way of looking at the subject. There are text-only communications packages, graphics-only packages, and graphics-and-text packages.

Netweavers should feel free to call any software that handles *any* kind of outside communications *netware*. Since the trend is toward more and more "integrated" functions in single software packages, current distinctions between categories of netware are rapidly blurring. Most communications software up to now have been "stand-alone" packages. That is, they perform a single set of related functions, and no others. They do not generally create files that can interact with other programs.

Integrated packages, on the other hand, can ostensibly do a number of things: word processing, communications, spreadsheets, and graphs. There are advantages and drawbacks to this approach.

4.14.1   Integrated Netware Systems

Examples of stand-alone, off-the-shelf packages include WordStar and VisiCalc, to name two that have enjoyed widespread use. Then there are so-called software families, wherein each package in the group, even though operated on a stand-alone basis, creates files of data that can be accessed by any other in the same family.

The advantages of the software family approach are obvious. Since the programs can exchange data with one another, users can manipulate the same information in many different ways with no need to reboot (restart) a program, do constant diskette swapping, data reentry, utilities conversions, and so on. A family may also be easier to learn over the long run, since the programs making up the group often use similar or identical commands and procedures. The last advantage is economic: startup costs can be kept down. You buy only the initial application(s) you need, with growth/expansion possible later.

On the negative side, different kinds of software authors are good at different things. A family produced by the same group of software writers may end up supplying a great word processor but a lousy communications package, or vice versa. Software authors generally are not consistently good at all aspects of software writing. Writing software is still essentially an art. Like the good movie directors who mellow with age and experience, software writers also get better with experience, turning out ever-more-exquisite computer solutions. But they may also produce duds along the way.

The integration approach to software—doing more in a single package—suffers from the same drawbacks as software families. The same advantages also apply. In addition, while the initial price of an integrated package is usually much higher than that of a stand-alone package, there may still be an overall price advantage when you add up all the separate functions and figure out what the same set of functions purchased one at a time would cost.

A *communicating word processor* is an example of an integrated netware product. You can create files with the word processor, consisting of whatever kind of documentary text material you choose, and then use the communications section of the same package to transmit the text files to your mainframe or your typesetter, or whoever. Again, the advantage is that you don't have to go through extra steps (changing diskettes, reloading a different program, perhaps converting a file from one format to another) to go from editing to transmitting the document.

If you don't mind the extra steps, though, there are many products available off the shelf that *do* work well with each other, even though not designed as a family or even written by the same authors. In many cases, this is the way to go for several reasons.

1. Off-the-shelf packages that have been on the market for some time are pretty thoroughly debugged. You won't have as many nasty surprises waiting for you as with untested integrated packages.

2. You can select each package according to the best possible criteria—according to *your* lights—for that function, rather than having to take potluck on a family of features you may not want, need, or even like.

3. Many stand-alone packages are "freeware"—in the public domain as far as copyright is concerned—and have very good price/feature ratios.

4.14.2 Terminal Netware

Programs that turn a general purpose microcomputer into a functional look-alike of a much less powerful terminal—dumb, smart, or otherwise— are called terminal programs. Sometimes this is a wise choice, since specific mainframes may want the micro to look like a specific brand or configuration of terminal. The appropriate software does this.

Things to look for generally in terminal netware are the following.

- Does the program provide a space in RAM, called a *buffer,* for the temporary storage of incoming data? Will that space be large enough for the files you are intending to exchange?

- Are all the protocols you will need provided? If you are going to be communicating with many different kinds of systems, are the protocols adjustable within the software? Are the protocols supported proprietary? That is, will the micro or computer at the other end have to have a communications package identical to the one you are running, or have to get a similar package from the same supplier?

- Does the program have so-called *macro* capabilities? Macros let you define and store log-on sequences and telephone numbers on diskette. They can then be recalled and used with single keystroke commands.

- Does the program allow you to switch easily between communicating with a remote computer and communicating with your own micro? This feature allows you to change protocols "on the wing," make minor editing adjustments to files, and do other chores while remaining on-line and connected to the remote computer. When you have accomplished the local task, it should be easy to switch back to communicating with the outside.

- Does the terminal program allow you to transmit files made by the word processor(s) you currently use? Is there a file length limit? If so, is it large enough?

- Can the program take full advantage of all the micro and modem features you have? For example, if the program doesn't support 1,200-baud communications, you won't get full use out of the 1,200-baud modem you intend to use with it.
- Are the producers of the program available to help you with it if you get into problem areas?
- Does the program support the data security features you need, such as encryption or passwords on sensitive files?

4.14.3   Smarter Netware

The more a given communications program can do for you, the easier you and your fellow networkers will have it. The more the netware can do, the smarter it is said to be. Smarts cost money. Smart netware will cost at least as much as good word processing software, and will cover the same range of prices as well. Again, there are many good terminal programs in the public domain, or even inexpensively through users' groups. Some of these programs are even available over the phone lines themselves. Catch-22 with these packages, though, is that you have to be able to dial up and hook into another computer to get these programs in the first place.

4.14.4   Micro-to-Mainframe Netware

Some programs are specially designed to communicate with specific kinds of mainframe computers. Such programs are usually also called *terminal emulators,* since they make your micro mimic the communications behavior of other brands of terminals. This is a very specialized type of program, as you might suspect. Many of them have high prices that reflect the degree of specialization involved. Even with the special software, an additional piece of black box hardware may also be needed to get your micro to talk to a given mainframe. This black box may go at your micro end of the link, or it may be needed at the mainframe end.

4.14.5   E-Mail and CBMS Netware

If the majority of your communications are or will be one-to-one, private, and/or personal messages, then you might want to look into electronic mail (e-mail) netware. Sometimes e-mail services are called computer-based message systems, or just CBMS. Like most acronyms in the emerging information industry, CBMS can refer to a wide variety of systems, including computer conferencing.

Most e-mail systems allow you to dial into a preselected computer and send your message to a specific user of that system. Some packages allow you to send Western Union telegrams and telex messages. Some are built into special modems, featuring timer and autodial functions that will send your "mail" automatically.

Electronic mail has many advantages and only a few disadvantages when compared with regular mail via traditional postal services. The main advantages are:

- Faster delivery
- Reduced expense over paper-based mail (photocopying, postage, etc.)
- Geographic independence
- Time-zone independence
- Improved access
- Reduction in internal delivery system overheads
- Improved coordination of group tasks
- Shorter messages, hence shorter processing time
- Automatic generation of records
- Reduction of expensive hardcopy file spaces
- Better search and retrieval
- Independence from weather conditions and/or public holidays
- Reduction of tendency to play "telephone tag" with colleagues

Disadvantages of electronic mail include:

- Reduced control over input (faster junk messages)
- Pressure to respond instantly
- Increased message traffic (it's cheap, so why not send the message?)
- Tendency to use it even when other communications channels may be more appropriate to the situation
- Tendency to generate and keep records that may be useless in the long run, i.e., garbage files

Although e-mail looks pretty good on paper, there are several factors to keep in mind before charging out and buying into a service, software, or hardware just for its proffered advantages.

Most of the mass-oriented timesharing services, such as Compu-Serve and the Source, offer some form of e-mail so that you can send electronic messages to another person. That person, of course, must also subscribe to the service. If she doesn't, then you have to find some other way to send the person e-mail. The U.S. Postal Service offers an electronic messaging service similar to that provided by Western Union's Mailgram service, but, again, that service is not universally available yet.

The best way to approach the question of e-mail is to look at your current message patterns. While it is true that you can save money using e-mail in many cases, it is also true that the savings realized on each individual message can be quickly eaten up by service charges and connect fees. Sometimes a good solution is to install your own dedicated micro to serve as a central e-mail or computer conferencing system that can handle the e-mail needs of your company, group, or micro-owning friendship network.

Another problem is transmitting messages across networks. Say you are a subscriber on XYZ Network and your associate in Metropolis is a subscriber to CompuServe. Should one of you switch so that you can send e-mail messages? Not necessarily. The concept of gateways and bridges, not to mention black boxes in hardware and software, is beginning to permeate the telecommunications community. This means that more and more services are providing their users access to other services. Eventually, intercommunication between networks will be no problem at all. For now, the wise netweaver will shop and compare, prepare cost estimates and user traffic estimates carefully, and then decide on how best to implement e-mail or some other form of CBMS.

Finally, there is a major unsolved question in the use of e-mail services: the question of privacy and security [3.3.4, 4.10.4]. In the past, GTE's Telemail service, the Source, and other services have had files rifled by "unauthorized" people, including system operators, adventurous hobbyists, and government agencies. Even the sacred U.S. Postal Service is subjected to unauthorized breaches of privacy insofar as those who actually run the service can read messages as they pass through the system. As one observer put it, it's like leaving the keys to the office and file cabinets sitting around for anybody to pick up and use.

Electronic mail is not protected from government intrusion the way ordinary paper-based first-class mail is. A warrant is required to open private first-class mail, whereas electronic mail may be made available to officials on demand. This bodes ill for the future of our First Amendment rights.

As long as electronic mail systems potentially or actually create centralized records of who writes what and to whom, they will be targets for government and private surveillance. They also bring the likelihood of investigative fishing expeditions, political manipulation, and search-and-seizure without due process that much closer. Western Union's Mailgram service, for instance, stores copies of each piece of e-mail passing through the system—some 40 to 50 million messages a year—for at least six months. When law enforcement officials, perhaps under political motivation, present a subpoena to Western Union, the company's computers can scan the storage bin of messages for names, addresses, or topical keywords.

The solution? Demand a public key encryption system [4.10.4] from your electronic mail service, for starters.

4.14.6   BBS Netware

Software that turns your softspace into a host system or host node for an entire network of users comes in a lot of varieties. Some of the packages available in the above categories will also auto-answer your incoming phone calls and allow a user to send you information you may then access locally at your convenience. Bulletin Board Systems (BBS) are designed to host a different kind of communication and allow users to create their own information structures and contents.

BBS netware connects with the call-in computer, terminal, or micro, puts the user through a log-in sequence of some kind, and handles the protocol and housekeeping tasks associated with collecting and storing the messages of the users. The log-in sequence may or may not involve questions and answers about where the user lives, what your name is, what the passwords are, your mother's maiden name, and similar kinds of nonsense. Most BBS systems are experienced just the way you might experience a real bulletin board at your local self-service laundry or supermarket—assuming you are familiar with such establishments.

4.14.7   Conferencing Netware

Computer or electronic conferencing is the latest generation in netware development. Such software is oriented toward flexibility and ease of use, general purpose communications allowing both individual/private interaction and group interaction. Features to look for include the following.

Short training times. Users should be able to start accomplishing work on the system within fifteen minutes' exposure MAX.

Robustness. The system anticipates and filters out possible user abuse or error, without crashing or going off-line.

Advanced features. The system's capabilities should exceed the user's current level of expertise and use, yet not force the learning of advanced features. This might be called simplified sophistication.

Anonymity and privacy. All telecommunications programs should support both anonymity and privacy modes. Any software that makes it easy for the system operator or other users to monitor anyone else's use should be avoided.

Housekeeping ease. The system should be easy to maintain and keep current.

Adequate size. Memory storage, number of users supported, and communications speeds should all be adequate to the actual traffic patterns that you'll expect on your system.

Flexibility. The system operator and users should be able to adapt the software to a wide variety of communications needs, from order entry to opinion polling, from voting to small-group conferences.

_____ 4.15 NETWEAVING: GOING ON-LINE

Let's assume it's all in place. Your netweaving experience is well begun, but not nearly half done, as the proverb would have it. At some point, you will find yourself "flying solo" through the telecommunications web-works. This includes the mastery of logging on to many different kinds of systems with many different kinds of log-on procedures, mastery of the telephone services at your command, and at least a degree of mastery over your micro and its local attachments. All of this is well within the grasp of those ready to reap the info-rewards of all this mastery. Here's a last-minute checklist you can use before going on-line the first time.

- Record your communications parameters and the preferred parameters of the system(s) you will be using. Sometimes this is not possible, and you might have to experiment on-line. This is expensive, so be sure to record the results.
- Get notepaper and pen ready. You will want to document your connection procedures yourself the first time in any case.
- Check your local connections: cables, modem to phone lines, modem to computer, etc.
- Check your printer and get it ready. You may want hardcopy, and you don't want to take valuable on-line time to ready your printer or feed it paper.

If you go through all this and still have difficulty, here is an additional general checklist you can use for troubleshooting while on-line.

- Are your protocols set correctly? Parity? Baud rate? Full or half duplex? Character length (data bits)? Stop and/or start bits?
- Most timesharing services allow you to specify what sort of terminal or micro you are using and what the line width of your display is. If your on-line display looks bad because words are getting broken at

the ends of lines, check your manuals and find out how to tell the transmitting system what your line length is. Or call your customer service rep for the same data.

- Are the modems set correctly: one as the "originate" and the other as "answer"?
- Is the modem connected properly and powered up?

4.16   LOCAL AREA NETWORKS (LANs)

So far in this chapter, we've been dealing with the technical foundations and concepts needed for individual use of telecommunications and netweaving. This foundation can be used to good advantage as the netweaver "scales up" her system, adding capabilities as newer, lower-cost technologies come along.

One of the first things a netweaver in a small business or corporation might want to do is create a local area network, or LAN. A LAN is usually found in the same building or building complex. Because it uses its own wiring, and because the distances involved are usually much shorter than those associated with the use of the voice telephone lines, a LAN can support higher baud rates.

A primary consideration for a LAN is flexibility. It should be able to accommodate new nodes without interrupting the ongoing operation of the network. It should also be designed so that failure of one node cannot bring down the whole network.

The three resources to be considered by a netweaver wishing to select and design a LAN from among the available options are micro time, peripheral time, and human time. The most expensive, of course, is the time of human beings on the network. This fact can often justify the apparently greater expense of high-speed peripherals and high-performance LANs. By making device-sharing possible, a local network can reduce the costs associated with providing expensive high-performance printers or hard disks to each individual micro user. The LAN has an added organizational advantage in that it links users who may then communicate with one another as well as with peripherals and other networks.

The LAN marketplace has been described as "chaotic" and "anarchistic." So many companies, large and small, have entered the so-called office automation arena that selecting the right mix of technologies for a local network is a much more formidable proposition than simply outfitting a personal softspace. Some of this chaos is inherent in the technologies themselves: there are many different ways to design an information system.

The almost sudden interest in LANs on the part of companies and the general public has arisen from several factors. The installation of a large number of individual micros is one factor. Many people now want to share files or upgrade to a hard disk system or better printer, and are looking for ways to cost-justify or cost-reduce their wants. This might be called the *end-user factor*.

Another factor, as usual, is that circuitry and designs are available now that weren't available as little as six, three, or even two months ago. This is the *technological factor*.

The third, but probably not the last, factor pushing the LAN wave is the much-touted *productivity problem*. Even though formal studies have shown that the typical executive or manager's job is difficult to automate, the push is still on to increase the throughput of the typical office. In most cases, this means automating clerical and information or data management functions. However, much of the sales appeal of LAN technology comes from its promise to make the typical executive, whether manager or mere bureaucrat, more "productive." Never mind that "productivity" among this class of workers cannot be adequately defined or measured.

As a result of all these factors, there are many competing LAN technologies. Just which LAN technology will predominate over the coming years is still an open question. It is perhaps a less important question than asking: Which one currently represents a cost-effective solution to my LAN netweaving problem? The technology that will be available two, three, or five years from now will be so different from today's that many netweavers may find it economical to scrap everything and start fresh several times before then.

A direct connection between two micros in the same location, through their serial ports, using a crossover cable or two modems, is essentially a simple kind of LAN. The need for making this kind of connection arises over and over again in the micro world. Sooner or later, a netweaver is going to want to transfer files between two machines. Sometimes they will be dissimilar-brand micros, and connecting them will tax your knowledge and ingenuity. Many times, however, they will be identical machines. That's a little easier. Be methodical and scientific in your approach. In addition to keeping notes on the various systems and interconnections you experience or make, you will want to observe the following netweaver's guidelines.

- Are you using a null modem cable [4.13.2 and Appendix IV]? If so, is the wiring correct?
- Always keep one side of the communications link "constant" or "steady state." In the case of remote time-sharing or BBS systems, that's simple enough. In the case of locally linked micros, set one system's parameters (record them) and then leave them alone while you twiddle with the other.

- If the two micros are connected directly using a modem on each, is one modem set up as the "answer" and the other as the "originate" modem?

- Change only one thing at a time, try it out, and make a record of the results.

- Persevere. You are dealing with a multivariable system. The combinations and possibilities are numerous but not infinite. With patience and time, you will solve any given problem in communications.

_____4.17   LAN CLASSIFICATIONS

LANs can be classified in several ways: according to their speed or *throughput,* according to the protocols they use, according to the method of their interconnection, and, finally, according to their topologies. There are probably umpteen other ways to classify LANs as well. This proliferation shows no sign of abating, so let's just plunge in and look at LANs from each frame of reference.

It must be pointed out here that the overview as presented in the following sections necessarily glosses over a lot of technical detail and technical disagreement among experts who work with and write about the emerging world of LANs. There is much less agreement at the LAN level than in the rest of the telecommunications universe.

4.17.1   LAN Protocols

Technically, a LAN design must involve the same factors we looked at in sections [4.5] through [4.15]. Perhaps the most hotly contested design area involves the adoption of standards for LANs. By looking briefly at a few of the major competing protocols, the netweaver can get a sense for what's involved. In the end, the protocols used may or may not affect the flexibility of your LAN, but knowing a bit about them (pun intended) can aid in the design and selection process.

There are two network control protocols in use. Network control has to do with how each machine gets access to the network and its assets, such as a printer or modem. Since there are usually more than two machines in the network, some means has to be devised to give each of them access in a way that is fair.

With the *token passing* control method, a marker is passed from machine to machine. The node with the token gets to speak. This is a little like the conch shell symbol of power in *Lord of the Flies.* In that novel about English schoolboys marooned on a desert island, the conch shell bestowed power to speak on the boy who held it. The rest of the group had to shut up until the conch was passed.

The CSMA/CD (which is a gawdawful acronym standing for carrier sense multiple access with collision detect . . . aren't engineers wonderful?) method of network control is more polite. Each node listens to the network channel. If no one else is using the net, then anyone may speak. If two try to get the net at the same time, the *collision* is detected and they both shut up, wait a few microseconds, and then try again. The amount of time each waits before trying again is variable insofar as more or less random intervals are chosen by each node.

4.17.1.1
IP/TCP

This is the Internet Protocol/Transmission Control Protocol, set and used by the U.S. Department of Defense since 1980. It is oriented toward peer-to-peer communications. It is important only to the military, and those wishing to interface with the military.

4.17.1.2
ETHERNET/XNS

Xerox Networking Systems protocols are being used by many independent suppliers of LANs, including Nestar and Davong, who make hard disk drives, and 3Com, a peripheral manufacturer specializing in micro communications. This is the Ethernet standard. This standard has significant cost advantages for one hundred or more users.

The prime virtue of Ethernet, according to engineers, is its simplicity (as compared, say, with IBM's SNA). A coaxial cable bus, with terminators at each end, is laid out. Stations can be added to the bus at any convenient point. Its speed is a major advantage: 10Mb/s. Ethernet uses no central controlling machine. A user gets on the network according to statistical arbitration. CSMA/CD is the controlling method used. This method is *load-dependent:* the greater the number of nodes on the network, the greater the access-delay times will be.

4.17.1.3
SNA

Systems Network Architecture is a proprietary IBM protocol. It may be necessary for a netweaver to deal with this protocol at some point, especially in a setting that already has many IBM computers, micro and mainframe, installed. The protocol is, in the words of old hands, "awkward and overcomplicated." SNA specifies the relationship between two other IBM specs: its "virtual telecommunications access method (VTAM)" and its "network control program (NCP/VS)."

4.17.1.4
ECMA

The European Computer Manufacturers Association (ECMA) protocols are being developed in cooperation with the International Standards Organization (ISO).

4.17.1.5
ARCNET

ARCnet technology is also a coaxial cable system, but its topology can be arbitrary. Node connectors called "hubs" are placed throughout the network. Many nodes can be connected to a single hub. Each node has its own "private line" to the hub. ARCnet operates at 2.5Mb/s. Access to ARCnet is controlled by token passing. Delay times for access are predictable, and the maximum access time can be known from the start.

4.17.2   LAN Wiring Techniques and Connections

Local area networks can be connected in many ways. There are four, though, that have come to be used more than others. In ascending order of cost and complexity, they are:

Bus
Twisted pair
Coaxial: baseband
Coaxial: broadband

4.17.2.1
BUS

Each node is connected using a flat cable, similar to the one that connects a printer with a micro. Buses use parallel transmission [4.7]. When a node wants to get the network's attention, it simply asks for it. If the line is already in use, the calling node must wait for the line to clear. This is like an old-fashioned party telephone system and is the main disadvantage of bus LANs. The main advantage is that the bus itself could care less what's hung on it. Tinsel, TRS-80s, Apples, or IBMs, it's all the same to Mr. Bus. However, even though different micro makes can be put on the bus to share things like printers or modems, the different makes won't *necessarily* be able to talk to *each other* on line.

Buses (as anyone who has had to wait for one knows) are rather slow. But this LAN solution is the least expensive of the current alternatives.

4.17.2.2
TWISTED PAIR

"Twisted pair" does not refer to Bonnie and Clyde. Rather, the "pair" is a pair of wires, and they are twisted together and shielded in order to minimize external electromagnetic noise from stray signals in the environment. The Corvus and Nestar hard-disk LANs use this kind of connection. Baud rates on twisted pair networks can get pretty high: close to 1Mb/s. This rate is sufficient for relatively small LANs subjected to intermittent use by its user group.

The main limitation on twisted pair LANs is distance. Nodes must be less than 2,000 feet apart from each other. Greater distances require some other transmission technique, such as a phone line.

4.17.2.3
COAXIAL:
BASEBAND

Now we're getting somewhere. But, as usual, we're spending a lot of money to get there. Speed and distance limits are expanded using coaxial LAN connections. Nodes can be up to a mile away from each other and can transmit info at the rate of 10Mb/s. Xerox's Ethernet uses this approach. It requires installation of a coaxial cable, which itself is somewhat more expensive than other cable types. The LAN can link as many devices as required. Again, only one user at a time can access the LAN. "Baseband" means, in effect, that there is only one channel available for allocation to a particular user. This is not as great a drawback as one might think, since the speed of the network assures that, for the most part, users will not experience any noticeable delays. Even though only one user at a time can access the net, the speed of the process makes it *appear* that everyone is on at the same time.

4.17.2.4
COAXIAL:
BROADBAND

Here the wiring is similar to that of baseband LANs, but more than one channel is provided to service users. Many users and high speeds characterize broadband LANs. This is expensive right now. This method allows voice, data, and video signals to be mixed on the same LAN, which is why cable TV stations are getting involved in digital data transmission and videotext. They already have the broadband cable in place. These kinds of LANs, while theoretically exciting and worth monitoring in the near-term future, will probably be out of reach of most netweavers for some time to come.

Both broadband and baseband coax technologies are highly reliable and well proven in the cable television industry (CATV), where equipment has been designed to hang for years on telephone poles under all kinds of weather conditions without showing the strain.

Many LANs can exist at the same time at different frequencies on a coax cable. Using gateways, the different LANs using the cable can also communicate with each other. Several LANs might share the cost of cable in a building and thus cut down on the installation costs for each of them.

4.17.3 LAN Topologies

Each LAN has an associated topology. Its topology is simply a map showing the connections and nodes in the LAN. Any specific LAN can have a totally unique topology, of course. Three topologies show up so often in LANs that they have their own names: the bus, the star, and the ring. Topology, as you might have guessed, is loosely associated with wiring and connecting techniques. Practice has linked the bus topology with bus wiring, the star with the twisted pair, and the star or ring with both baseband and broadband techniques.

4.17.3.1
BUS

The simplest and least expensive of LANs, overall.

**4.17.3.2
STAR**   The network server or host node is at the center. All other devices are routed through this central computer. If that goes, crashola goes the LAN.

**4.17.3.3
RING**   This is sometimes also called the "daisy chain" LAN. (Who said techies are not sensitive?) This topology allows information to be passed from node to node until it reaches its proper destination. A parallel path, *around* each node, is also provided, just in case a particular node decides to blow up. This solves the major reliability problem associated with a star topology.

4.17.4   Network Servers

Images of butlers in black, carrying hors d'oeuvres bytes on trays. . . . A network server or network manager is a microcomputer or minicomputer that attends to the traffic direction functions of a LAN. Other names

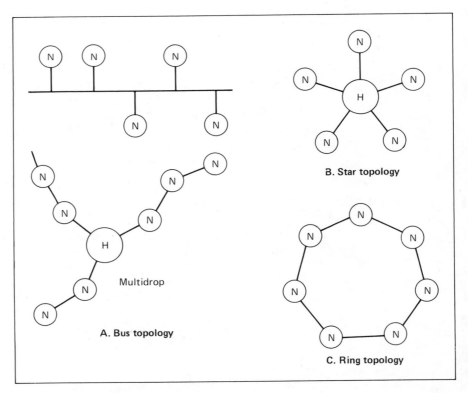

FIGURE 4.17.3. LAN Topologies

for this function are *host node* or *central node*. Whether or not your LAN needs such a function will bear greatly on its cost. That's because the host node requires a dedicated machine, doing nothing but managing the network and its traffic. There are LANs that do not require such a machine. The function of network butler may also be performed by some other peripheral or lower-cost black box.

4.18   LAN NOISE CONSIDERATIONS

Because they operate at high speeds, LANs are subject to interference from a variety of sources that can be found in the typical office. These noise sources are called EMI (for electromagnetic interference) and/or RFI (for radio frequency interference). Generally, the longer the wire connecting two devices, the more subject to signal loss and noise it will be. Coaxial cable systems are less subject to noise than other connection methods. The relationship between the length of the connecting cables in a LAN and the amount of noise that results has given rise to recommended practices in cable installation. The rule of thumb is that the lower the baud rate, the longer the cable can safely be. In any case, LAN cables should be installed well away from power line wiring (unless, of course, the LAN *uses* this wiring for its signals).

4.19   LANᴇᴛᴡᴀʀᴇ

There are some basic functions that most local area networks should be able to perform for its users. The networks require software to complement the hardware of the particular system. Such software may be required on each micro in the network, in addition to being part of the network server, if there is one. The basic functions are provided in programs called LAN utilities because they help to maximize the use of the LAN. They include at least the following.

- File serving: Permits users to share files on hard disk and with one another.
- Print serving: Permits users to share printers.
- Electronic mail: Permits letters, memos, and files to be transferred between nodes. May include a simple text editor.
- Gateway access: Permits users to send their output through a modem and direct the modem's operations.
- Computer conferencing: Permits users to create open files and ongoing dialogs in a common information space.

_____ 4.20 PBXs AND PABXs AS LANs

Many organizations and businesses have already installed in-house telephone switching systems. These systems are called private branch exchanges (PBX) or private automated branch exchanges (PABX). Since these systems are already installed, it is tempting to use the same wiring for data communications as well as voice. This means that a lot of businesses have built-in LANs already. Even though the speeds on such networks will usually be slower than the optimum possible with a dedicated LAN cable, they will remain attractive. New wiring costs a lot, and integrating voice and data on the same system can be quite convenient for overal communications needs.

By digitizing the voice communications of a PBX, the same terminal or work area can be used for both voice and data messages. Why digitize voice communications at all? Because it permits multiplexing [4.8.4] of many messages on trunk lines, which is a more efficient use of a scarce resource.

Long-distance communications account for most of the tab for business telephone use. Besides, voice communications predominate in most organizational settings. PBXs are potentially the least expensive way of handling the average office mix of voice and data.

PBXs can support more nodes or terminals than any other current LAN technology: up to 30,000 in some systems! The 1983 prices on PBX nodes ranged from $270 to $1,600 per installed station. Baseband or broadband LANs, in contrast, averaged $2,500 per station (for integrated voice and data).

_____ 4.21 BRIDGES AND GATEWAYS AND LINKS (OH, MY!)

Look at it this way: your modem on your personal micro is an electronic _bridge_ between your physical softspace and the outside world. It is part of the membrane through which information passes, in and out. There are bridges between networks, too. These _link_ two networks together, so that data can flow between them. Cities in which such links are located are sometimes called gateway cities. All information that needs to pass from network A into network B must be routed first to the gateway city, which is where the bridge is located.

A local area network may also have a bridge to the outside. This bridge may be narrow, having only one line out connected to any given inside node. Or it may be wide, allowing many users to pass through at the same time. The latter usually requires some kind of MUX [4.8.4] or MUX/modem combination.

_____4.22   THE LAN DESIGN PROCESS CHECKLIST

The following list of design questions and precautions will help you establish a LAN.

- Design the system on paper first. Don't connect anything with anything until you know exactly what you're doing.
- Rules of thumb: The higher the baud rate of the LAN, the shorter the distances involved will have to be. The longer the distances, the higher the cost. The greater the bandwidth, the higher the cost.
- Inventory your real requirements. How many stations or terminals will you need? What kind of peripherals are available now? What kinds will be needed to serve the group you have in mind, now and in the future? Will there be modem links to other LANs or outside services?
- Will you need data only? Or some combination of voice/data/video?
- If the number of terminals required is between three and eight, then look at multiuser microcomputer systems rather than LANs. This might be much less expensive. However, most of these are turtle-slow. Test before you buy.
- It may not be wise to pay for bandwidth you don't currently need and can't foresee needing in the near future. If most of the users of the LAN will be part-timers, then lower speeds may be tolerable because the LAN will likely never be used to the point where speed slowdowns are noticeable.
- In most LANs, the biggest bottleneck will be at the disk drive. Thus this peripheral will likely be the most limiting factor in the use of the LAN.
- What data do users need to share? Where will it be stored?
- Should there be an e-mail or conferencing facility built into the LAN utilities?
- Where will the users' terminals be located? It is helpful to draw a floorplan and plan cable lengths needed. These in turn will help in estimating final costs.
- Prepare a performance scenario that describes the overall system, with emphasis on the optimal performance desired now. The scenario should also describe what you think you'll need a year from now. A second scenario should be prepared describing the minimal acceptable requirements for now and a year from now.
- If all the parts of a LAN are to come from a single supplier, what software will run with it and be included with the system? What other software is available, from what sources and at what prices?

- Will you have to specify and contract for additional software (or have to write it yourself)?
- Is the LAN easily expanded? Can users be added to the system with no noticeable degradation of performance as more peripherals and people are added?
- After the paperwork design has been done, costs associated with the system can be assessed. These include cost of wiring, nodes, peripherals, additional switch boxes, network server hardware, hardware for each node (if required), software, and don't forget labor, including consultant's or programmer's fees.
- Draw up an implementation plan that includes milestones. On such-and-such a date we'll have the hard disk drive up and running. By this date all the terminals will be connected. Include training and support for your people in the plan.
- As with any computer-related project, it is wise to design the implementation phase so that vital functions are maintained during the switchover and shakedown periods. *Never* transfer a currently workable manual or semi-automated system over to a new, untried system before a thorough testing and debugging phase. This will probably mean dual systems for a time, which some cost-only managers may object to, but the extra cost involved should be seen as a premium on disaster insurance.
- Until micros and LANs become *much* easier to use, training and support from vendors will remain a critical factor in the success of a given LAN, especially those beyond a micro-to-micro linkup. As with any highly sociotechnical process, it is probably wise to find a previous client or two who have actual experience with the LAN you want to weave.

4.23   SOFTSPACE EXPANSION AND SCALING UP

Someone—I don't know who—once defined freedom as "the widest possible range of choice and options." Some decisions inherently close off options in certain arenas. Others tend to widen the choices in one arena while narrowing them in others. Some decisions are inherently open-ended—the decision to engage in continuous learning, for example—and widen horizons beyond what we might be able to imagine.

In the netweaver's chosen arena, the word "option" is the prime operator. Even though some specific technical decisions may hamper you in the short run, in the long run your personal and organizational freedoms will increase, no matter what your entry point into the Information Transformation happened to be.

Oh, it *will* require work, to be sure. Changes in self, group, or organization don't occur all by themselves. But as time goes on, the number of ways in which you'll be able to enhance your freedom of information will tend to multiply.

The options that follow are only a beginning.

4.23.1   The Paper Connection

Far from eliminating paper altogether, the information economy has spawned whole new reasons for paper and ways of using it. It has also eliminated the need to use paper for certain things, such as the long-term archival storage of documents.

It seems likely that, especially as on-line graphics get better, small-circulation newsletters will go from paper to video screen sooner than daily newspapers. There are many ways in which the telecommunications link can speed the delivery of paper-based information. Electronic publishing can include streamlined production, typesetting, and layout of magazines, books, and advertising material.

The simplest way from machine-readable to paper-based is through a low-cost printer installed in your primary softspace. If you begin with just a (paper) printing terminal to access existing networks, you'll begin to get results in the form of information you can capture on-line and begin to use after you're off-line. For safety's sake, you'll want to make hardcopy printouts of some files and archive them separately from your floppy disk storage and primary workspaces. Of course, if you have access to microfiche or microfilm technology, the long-term storage of such material becomes less problematic.

Going the other way around, from paper-based to machine-readable, is becoming easier as well. Early-generation reading machines cost tens of thousands of dollars. They are now down to thousands. Before long, an input device capable of reading printed matter from books and magazines will be available for connecting to your micro. Then you'd better have at least a hundred megabytes' worth of hard disk storage, however.

In the matter of newsletters, journals, and the like, two routes to paper are becoming increasingly widespread, and neither shows any signs of giving way to the other. The first is the printed newsletter created wholly or in part on-line, by a user community conferencing or on-line journal group. Printouts are made on low-cost dot matrix or letter-quality printers, and reproduce through offset printing. The Apple Macintosh with its ability to print varied graphics on a dot matrix printer is a delight for groups wanting to produce low-cost yet attractive newsletters.

The second route to print is through a typesetter, regardless of where or how the original copy has been created. The typesetting company takes copy on-line from a transmitting microcomputer. Or floppy disks are sent to the typesetter, who uses the files therefrom to go directly into the typesetting machine.

Converting your draft of a report or manual or book directly from your word processing disk to type can save a lot of money. A typesetting firm's greatest cost associated with settting text in type is the human labor required in entering the information into a machine-readable form in the first place, and, associated with that, entering the required control codes that actually tell the phototypesetting equipment which type fonts to use on which letters, paragraphs, and headlines. These two functions—text entry and control—are separable only to a degree. To the degree that an author or company can provide both functions at the same time, to that degree will money be saved on typesetting costs.

In the best scenario, you should be able to give your data disk to the typesetting company and forget about it. To go a step further, you should be able to send your text to the typesetting firm on the phone network and then forget about it. Alas, netweaver, computopia in the typesetting arena is only in the beginning stages. Still, that's better than nothing.

Many typesetting firms are gearing up to provide services to users of micro telecommunications networks and word processors. Some of these firms are associated with one or more of the mass-market-oriented data base services such as CompuServe. Our best advice in the matter is to find a local firm offering these services. Don't use a mass utility if you don't have to, because these intermediaries can eat up any potential dollar savings you can achieve by going this route. Don't even think about typesetting a large document this way unless you either can give the company your data disks (via U.S. mail or other delivery service) or have at least 1,200-baud communications ability from your softspace. There will be no savings at all if you are transmitting a lot of material at 300 baud over phone lines.

The communications software you use for this purpose will also be critical to your success. The terminal software should be able to create translation tables [Appendix IV: 4.7.2] that take your micro's control codes and translate them into codes understandable by the typesetting machine at the other end. Your typesetting shop should be able to provide you with these tables. The shop may even be able to recommend a preferred communications program and word processor combination that works well with the particular typesetting machine involved. Your terminal program should also support the "X-on/X-off" (control-Q/control-S, ASCII [Appendix IV: 4.7.2.1]) protocol. It is also essential that your communications package and modem combination are capable of sending the 128 standard ASCII and 128 so-called *extended* ASCII characters. The IBM PC, for example, uses an extended ASCII. Most packages support the ASCII codes from 0-127, which includes alphanumerics and some control codes. When working with typesetting machines, a greater range of codes is desirable, if not mandatory.

For on-line journals and graphics, the NAPLP protocol promises to make even small micros capable of displaying adequate graphics retrieved from remote data bases. Of course, to print out such graphics will require an appropriate hardcopy printer.

4.23.2   Cellular Mobility

The use of mobile phones in automobiles has always been associated with mile-long Cadillac limousines. No more. The advent of digitally based cellular mobile radio ensures that mobile phone antennas will be sprouting on Japanese compacts in no time at all. The initial appeal of cellular radiophones will be to those who are on call in highly sensitive and time-sensitive professions: doctors, maintenance technicians, diplomats, and stockbrokers. That has been the "traditional" market for mobile phones and pocket pagers or beepers. However, the advantages of being able to conduct business while stuck in a traffic jam (before telecommuting takes over enough to eliminate such beasts) will be appealing to many.

The concepts of the pocket pager, the mobile telephone, and the portable computer or terminal are fast merging. Devices are already on the market that enable you to contact your home computer from any telephone and operate it remotely. There are also devices that will digitally store and forward written messages to you which you receive on a pocket device with a small readout. You can set the device to notify you with a beep when a message comes in. The same principle can be applied to transmitting continuous data and updates to you for following the stock market—or the continuously monitored results of a scientific experiment.

Figure 4.23.2 shows the main components of a cellular mobile radiophone system.

4.23.3   Digital Amateur Radio and Amateur Packet Networks

Eventually, citizens band will "go digital" so that anyone with a CB set will be able to use the CB bands to link digital devices. If these bands are "packetized," then much more efficient use will result. Work in this direction is already taking place among the amateur (ham) radio community. All the techniques of digital packet transmission can be applied in the amateur radio bands.

Hams have been pioneering low-cost and public domain protocols in hardware and software for some years now. They have been digitizing video images and experimenting with slow-scan television (SSTV) for an equally long time. They have been experimenting with voice packetizing as well.

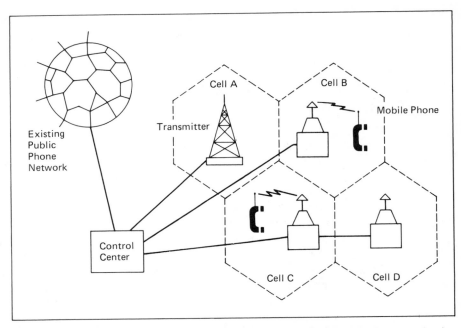

FIGURE 4.23.2. Cellular Radio. Cellular radio combines digital, radio and telephone technologies to greatly increase the supply of mobile communications channels. A given city is divided up into equal geographic regions ("cells"). Each cell has its own transmitter. As a user with a mobile phone moves from cell to cell within the city, her calls are automatically routed through the control center computer to the appropriate cell. By using digital signals combined with low transmitter power, a given radiotelephone frequency can be used more effectively, carrying more calls than if a single, centralized and high-powered transmitter were used.

4.23.4   Video

As effective bandwidths increase and storage gets cheaper, video will get integrated with voice and data on the same networks and in the same devices. Although few pieces of hardware are available to the individual consumer as yet, the trends are already clear. Video is going "component" and "digital" at the high (expensive) end already.

The creation of a personal medianetwork that allows a personal micro, video recorder (VCR), and component television system to work together with a vast amount of digital storage and the phone networks will mean that the consumer of information can also become a producer of information. The investment required for such integrated systems is currently quite high, but not as high as it used to be nor as low as it will become.

4.23.5   Fiber Optics

Fiber optic cable systems may be the ultimate netweaving tool. Optical fibers are tiny glass "pipes" through which light beams are transmitted. Information is encoded in the light. The advantages to optical fiber are its practically unlimited bandwidth and transmission speeds. Currently, the cost for optical fiber technology is too high to make it practical in all but the largest of applications. But as the Information Age gathers momentum, demand for better conduits, faster speeds, and the ability to handle high-quality video images along with our computer programs and text will ensure that the fibers begin growing in earnest.

———— 4.24   PLANETARY LINKAGE

Today there are many barriers to international digital communication between individuals. The creation of a personal global network, in other words, is hampered by nationalistic political forces. Banks and corporations are having an equally difficult time getting their networks in place internationally, but these organizations can afford to pay off governments to allow specialized placement of dedicated equipment and channels.

Truly global egalitarian networks or network-based organizations will not be possible until anyone, anywhere, can dial up anyone else anywhere else with no more thought than dialing down the street to the local butcher. Technically, this is possible today. But given the state of telecommunications in other countries, even the wealthy have difficulty communicating by transnational voice phone today. Before widespread digital panmedia networks can emerge, the technologies and configurations of all the telephone systems in the world must be improved. Unless . . .

Unless a satellite transponder or two or more are dedicated to upgrading *everyone's* communication, not just the banks' and national governments'.

The portable, personal communications device integrating voice, data, and pictures is not far away. The broadcast power and sensitivity of geosynchronous communications satellites are increasing as the space shuttle moves larger and larger payloads into orbit. Today's space shuttle can put a 65,000-pound payload into orbit. (That's 355 times the size of *Sputnik,* dahlink.) Soon the ability to broadcast a signal that can be decoded by a small receiver carried by an individual will make global access possible for *anyone who can afford the device.* If the market is *global,* there is no reason to suppose that such a device cannot be mass-produced and distributed and thus brought within the reach of even the poorest of Third World information have-nots.

But then, global events and systems are not yet, by any stretch of the imagination, "reasonable." Catch-22 is that global systems that result in the reduction of the arms race and the cessation of genocide and ecocide are not likely to come about in the absence of a universal global access that just might result in a more reasonable outcome for planetary systems. Perhaps there will be a new "pioneering country" that makes this one of its goals despite what other countries may say or do to oppose it. Any bets on Japan? New Zealand?

_____ 4.25   FUTURETECH?

Digital integration—doing more with more media with fewer physical parts—is a given. Community media centers, combining computer database access, copying machines, telex and TWX and telegram access, satellite downlink channel access, and telephone access will begin to spring up first. Then the capabilities of these centers will radiate outward into the homes and offices of individuals.

Teleport dwellings—offices and homes using satellite downlinks and uplinks to all the networks—will begin to emerge, and all new architecture will incorporate integrated digital networks. A kind of homenet, if you will.

The use of space for manufacturing chips will result in further miniaturization and more complex chip design. This has also been a given in telecommunications for a long time, reinforced in the daily news.

The completion of "fifth-generation micro designs" will result, eventually, in the design of new micros by their predecessors. Again, this is already happening. One of the outcomes of this process will be the most general softspace capability of all: you'll be able to tell a factory what you want in an information device, and the factory will custom-make it for you, including any required new circuit design. It will be as cheap for the factory to make your unique device as for the same factory to mass-produce, say, a Walkman.

All appliances will become, willy-nilly, information appliances. The planet's ecology, technology, and information will merge. That which was last is now first. A perpetually new beginning in the Eternal Present. Omega.

APPENDIX I

Netweaving Development Tools

Huxley, Aldous. *Literature and Science*. New York: Harper & Row, 1963.

After C. P. Snow wrote *The Two Cultures*, alleging the existence of a "rift" between people of science and people of the humanities, Huxley wrote this reply. Not so much a rebuttal to Snow as it is an alternative way of viewing the languages employed by human beings in building models of experience, *Literature and Science* is a handbook, in its own right, for budding artists and scientists and those who would make so bold to as to be both.

1.1
LANGUAGE

Wilson, John. *Language and the Pursuit of Truth*. New York: Cambridge University Press, 1969.

Philosophy evolved from concerns of metaphysics to concerns of language and the right use of language, to the study of propositions, and, finally, to mathematics and symbolic logic. This slim introduction provides a clear understanding of what language is, what we can do with it, and how it produces meaning. Clear thinking has its roots in such understanding.

1.1.1
JARGON

The Jargon Game. Players sit in a circle. One player is the recorder. The group takes a few minutes to think up a jargon word that no one has heard before. The words and their definitions are written on 3 × 5 cards. (Teletrash, infomush, telesphere, microcosts, macrothought, selfware, freeware, footware, etc.) Players now take turns announcing their word to the group. Each member of the group tries to define the word. Player gets one point for each member who makes up an "incorrect" definition or passes. Players may import jargon words from other fields and make up new definitions if they wish. They may also use the same old definitions for the old jargon, but this is not encouraged. Players lose 10 points for each person in the group who has heard the old jargon word before and can give the correct definition. Player also loses 10 points when someone else "gets" the correct definition of the new jargon, based on what was written on the cards.

A variation on this is to make up a new game based on the new jargon created by the group.

This game has interesting applications in research. For example: What kinds of information are generated during a game of this kind? Is such a game playable on a telecommunications network? Does this game suggest any kinds of more rigorous (statistical, game-theoretical, group dynamic) experiments or games? Can machines be programmed to play this game? Would such a machine pass the Turing Test? Why or why not?

1.1.2
TECHNOTRANS-
LATOR

Lest the reader think I am playing the Jargon Game in this book, you should know that there is a company in San Francisco, Technology Translated, engaged in technotranslation. No sooner can you think up a whimsical bit of jargon than some Silicon Valley company has named itself that.

1.2
CYBERNETIC
ORGANISMS AND
ANTHROPOMOR-
PHISM

It's 1976. I'm on the beach in Santa Monica. Most of the people on the beach are in swimsuits and sungear. They are the dominant species on the beach that day. But what's this? A subspecies is traversing the sands. Scattered throughout the sunbathing, reclining, and sand-castling crowd are Senior Citizens, fully dressed, mostly male, but a couple of grannies, too—some even with hats!— wielding sticks with saucers on the end. The sticks have wires coming from them, going to headphones looking like safety earmuffs on the old ones' heads. Inexpensive metal detectors, it would seem, are the greatest thing for beach-combing since Hawaii. Metal detectors were originally "mine detectors." Now they are used for "mining" the sands of this rich beach. These miners are cyborgs: organisms with cybernetically extended senses. They are unique. They are socially and environmentally useful. Later, I was to observe newer cyborg species as the Sony Walkman phenomenon swept through the population. The life of an exopsychologist is never boring. Downlink hats will be all the rage in '89.

1.2.1
THE TURING TEST

Weizenbaum, Joseph. *Computer Power and Human Reason: From Judgment to Calculation.* San Francisco: W. H. Freeman and Co., 1976.

The major thrust of this work is that the world itself is being made too much into a computer. Weizenbaum claims that this remaking began long before the current popularity of microcomputers, before there were any electronic computers at all. Now we can use the metaphor of the computer to help us understand what we have been doing. In addition, Weizenbaum says, there are certain things we should not attempt to make the computer do, regardless of what we *can* make them do. This is certainly wise. What we should and should not make them do, however, lies in the realm of values and politics. These are realms in which technocrats are uncomfortable, if not oblivious. Hit them on the nose with a Weizenbaum. John Lilly has speculated that the whole *universe* may be one giant computer and us just itsy-bitsy bits doing some Supreme Programmer's bidding.

Evans, Christopher. *The Micro Millenium.* New York: Viking Press, 1979.

This is one of the few overviews of the development of the microchip that keeps its enthusiasm grounded in current and historical fact. Overall, Evans's view is optimistic without ignoring the real questions raised by widespread application of chip technology. An excellent introduction for cyberphobes, technophobes, neophobes, microphobes and phobophobes. As Evans explains the Turing Test (pp. 224–228), he points out that we judge whether or not a fellow human being is "intelligent" based on the signals he sends us. We infer everything from those signals, so why not use the same criteria to judge whether a machine is also intelligent?

Dick, Philip K. *VALIS*. New York: Bantam Books, 1981.

Metaphysical science fiction is the cutting edge of letters today. Dick defines VALIS (vast active living intelligence system) as "a perturbation in the reality field in which a spontaneous self-monitoring negentropic vortex is formed, tending progressively to subsume and incorporate its environment into arrangements of information. Characterized by quasi-consciousness, purpose, intelligence, growth and an armillary coherence." What's "quasi-consciousness"? Never mind. What's "mind"? No matter. Whatsa matter? Read *VALIS* and find out.

1.2.2
HEURISTICS

When you first encounter a field, you are innundated with Facts. Perhaps you try to remember all the Facts. This would not be productive in the long run. Although you may derive some conclusions from the facts in a formal way, it is far more profitable to identify and use a much smaller class of "critical ideas." For a good discussion of heuristic methods, see Campbell, *Grammatical Man*, pp. 220–221 [Appendix I:1.6.5].

1.2.3
THE CULT OF THE SACRED CYBORG

Vallee, Jacques. *The Network Revolution: Confessions of a Computer Scientist*. Berkeley: And/Or Press, 1983.

Definitely a cultish sort of book. Vallee lashes out at rampant anthropomorphism from the point of view of an old-time (mainframe) computer user and programmer. This book is less about networks and confessions than it is a nostalgia piece on the way telecommunications networks used to be. It is also useful for the negative examples it holds up for would-be netweavers. This one is essential for history-of-computing buffs. Everyone else may skip it.

1.3.1
WAVES OF CHANGE

Toffler, Alvin. *The Third Wave*. New York: William Morrow and Co., 1980.

Throw out two-thirds of this book and the netweaver will still have a wealth of ideas for new businesses to get into, trends to look for or help accelerate (or slow down), and insights that couldn't be gleaned without reading 1,001 other futurist books. The main criticism to be made of this work is that it is fragmented, disjointed, and unwhole. It is not written from a systems base, although it pays homage to the idea of systems. Its structure reflects nothing real, and it is therefore itself *très* Second Wave. But what an array of facts! And a lot of cute memes such as "indi-video," "practopia," and "indust-reality." And some not-so-cute ones like "informationalization"—ouch! Almost as bad as "compunications."

1.3.2
INFORMATION ECONOMY

Porat, Marc Uri. *The Information Economy: Definition and Measurement*. U.S. Department of Commerce, Office of Telecommunications, OT Special Publications 77–12(1), May 1977.

Dr. Porat has gone from academia to government to AT&T, and has helped that behemoth develop an Information Age advertising and P.R. strategy, which is something of a philosophy now among AT&T management. All based on Porat's rehashing of economic statistics showing that in 1890, 46 percent of the work force was engaged in farming, and only 4 percent in what has come to be defined post hoc as the "information sector." By 1980 these statistics were reversed: 4 percent of our population is now employed in agribusiness, and 46 percent move information.

Porat divided jobs into a "primary information sector" and a "secondary information sector." The primary sector covers anything that isn't intraorganizational in origin or use. For example, the people who produce books for sale are part of the primary sector. The people who produce an employee newsletter or write memos are part of the secondary sector.

1.4
DESIGN

Lieberman, Jethro K. *The Tyranny of the Experts: How Professionals Are Closing the Open Society.* New York: Walker and Co., 1970.

Lieberman argues, quite successfully, I think, that *everything* is much too important to be left in the hands of the "experts." Legal matters, health, buildings, accounting, plumbing, design of all kinds (down a list of some 300 occupations) must be taken back from the small groups who have staked out their exclusive rights to perform certain vital services. In medicine this is already happening to a large degree as "holistic health" puts more emphasis on individual responsibility for wellness. In the legal field, avoiding probate and do-it-yourself name changes and divorces are signs of the same trend. Telecommunications and microcomputers can serve to put more expertise in the hands of EveryPerson. Oh, and the last thing we want to do is create an elite corps of telecommunictions experts. Those who can control communications can control just about everything else.

1.5
MODELS

See Churchman [Appendix I:1.6.1].

Kuhn, Thomas S. *The Structure of Scientific Revolutions.* Chicago: University of Chicago Press, 1962.

When does a classic become a cliché? This book is cited so often that Kuhn has become like Freud and Marx: few people bother with the original source material. If you must cite Kuhn, netweaver, please read his essay. Or at least scan it. Normal science, he says, is a form of puzzle-solving. However, the puzzle only can consist of parts we already have in hand. The discovery of anomalous parts that "don't fit" into the current picture we have built tends to get ignored by normal scientists, because these anomalies are troublesome and psychologically distressing. In scientific revolutions, world views are changed radically. The same bundle of data is reorganized into a new system of relationships within a totally different model of the field.

1.5.1
METAPHORS

Samples, Bob. *The Metaphoric Mind: A Celebration of Creative Consciousness.* Reading, Mass.: Addison-Wesley, 1976.

Samples reminds us that all metaphors can be traced back to nature, which is different from culture. He encourages us to look through our inner telescopes and probe our brains and our atoms not as cultural voyeurs and mere number-gatherers, but as poets and children of the universe. This is a book about equilibrium and the taming of technology. Dance with it.

**1.5.2
MEMES AS
METAPHORS/
METAPHORS AS
MEMES**

Dawkins, Richard. *The Selfish Gene.* New York: Oxford University Press, 1976.
  A zoologist, Dawkins. He is yet another scientist who can communicate in letters. The concept of meme and its function in culture is Dawkins's contribution to our metaphor pool, excuse me, I mean our meme pool. The metaphor is slowly catching on and is quite useful in the Information Age, when information can alter information and computer programs can write other computer programs.

**1.5.3
ICONS**

Albarn, Keith, and Jenny Miall Smith. *Diagram: The Instrument of Thought.* London: Thames and Hudson, 1977.
  An iconic representation can merge with the diagram. In this comprehensively illustrated manual, Albarn and Smith present a method for applying the diagram to the internal world of thoughts, ideas, feelings, and emotions. Diagram your inner situation, they argue, and you will not only begin to understand your own attitudes and feelings, but you will also get more into touch with your innermost organizing mechanisms: spirit, soul, psyche, mind, biocomputer, or cerebral cortex. A diagram is a model of an idea. The nice thing about this book is that it is also an outstanding example of its own principles.

**1.5.4
GAMES**

Bell, Robert, and John Coplans. *Decisions, Decisions: Game Theory and You.* New York: W. W. Norton, 1976.
  A good introductory text on game theory. No one has yet designed a game that will teach things like game theory. In fact, the number of games that have yet to be designed to teach all manner of things is practically infinite. If you think you have a flair for teaching and a feeling for games, whole worlds are left to conquer using the micro, videodisks, and networks—combined with entertainment and distraction—to teach valuable subjects to thousands for $1,000s.

Davies, Hilary, trans. *Awareness Games: Personal Growth through Group Interaction.* New York: St. Martin's Press, 1975.
  This collection of games is a great resource for the netweaver who must work in small-group environments. The elements covered are introductions and getting acquainted; group formation and communication; observation, listening, and perception; identification and empathy; and aggression and self-assertion. These games are helpful in group formation and maintenance. Well-formed groups will accomplish their tasks more easily and efficiently. Switching from the "real-time" task to one of these games when the group is blocked is a useful strategy for keeping things moving.

**1.6
LIVING SYSTEMS**

Kenner, Hugh. *Bucky: A Guided Tour of Buckminster Fuller.* New York: William Morrow, 1973.
  There are many books by and about Fuller. This one is as good a place to begin as any. Netweavers would do well to familiarize themselves with the major concepts Bucky helped introduce. Synergy is one of those concepts.

Miller, James Grier. *Living Systems*. New York: McGraw-Hill, 1978.

A definitive work, drawing on many of the experimental sciences, rich in suggestions and implications for designers and educators. Since system theorists are by definition generalists, all specialists can benefit from having an overview of living systems as organized by Miller's model.

**1.6.1
SYSTEMS THEORY**

Bertalanffy, Ludwig von. *General Systems Theory*. New York: George Braziller, 1968.

Old Ludwig is the Grandaddy of systems folk, and his is one of those classic introductions in the field. It is more mathematically inclined than most of the other books on the subject. On the other hand, the later sections are excellent for their overview of the theory in such areas as communications and psychology. Humanists can ignore the first part of the book, dealing with differential equations and such, and concentrate on the latter half.

Von Bertalanffy explained the genocidal and amoral capacities of the human mind by pointing out that our societies have more values for complex organizations and systems than for personal conduct. This fundamental imbalance "dooms" individual morality because we are systems-within-systems and our values are increasingly context-dependent. This failure to comprehend systems, he says, has made our societies disordered and antagonistic. The antidote? Develop more self-organizing general systems of symbolic relationships to reconcile warring international systems and hostile academic disciplines.

Churchman, C. West. *The Systems Approach*. New York: Dell Publishing, 1968.
The Systems Approach and Its Enemies. New York: Basic Books, 1979.

Churchman manages to communicate the how of systems thinking better than most of the theorists writing on the subject. The 1968 text is still useful to the extent that it presents early basic concepts. It should be pointed out, though, that systems theory, like its subset cybernetics, has evolved considerably since this reference was published.

Churchman calls his 1979 work "just another step in the search for the meaning of *generality*, in this case a general design of social systems." For him, the "enemy" is the vast territory composed of social systems that have not yet been explored by what he calls "hard" systems analysts. He also points out that a large part of our social systemic reality is not any of the following: rational, irrational, objective, subjective, hierarchical, flat (nonhierarchical), and so on through a list of most of the major theoretical dichotomies popular today in general systems thinking. He reinforces my own bias that the academic/scientific "disciplines" are political, not scientific, categories and that one is not likely to learn much about systems by "dwelling in a discipline."

Weinberg, Gerald M. *An Introduction to General Systems Thinking*. New York: Wiley-Interscience, 1975.

This introductory text is a treasure trove of ideas that can be transferred into many other endeavors, including the so-called information industry. Among other things, Weinberg presents science, numbers, and the complexities of the world in a general systems light. General systems thinking is the simplification of science and the science of simplification.

1.6.2
CHANNEL AND
NET

See Miller [Appendix I: 1.6].

1.6.3
HOLONS AND
HIERARCHIES

Koestler, Arthur. *The Ghost in the Machine.* New York: Macmillan, 1967. *Janus.* London: Hutchinson, 1978.

Koestler, in *Ghost,* helped to dethrone behaviorism in psychology and helped make the study of human consciousness academically respectable again. He also helped make "holism" a buzzword of the '70s. In both books he brings out the usefulness of the holon as a concept. One of his "general properties of open hierarchical systems" is that parts and wholes, in an absolute sense, do not exist in the domain of living systems. The concept of the holon was intended to reconcile the reductionist-atomistic approach to living systems with the holistic approach. Essentially, Koestler was interested in alternatives to the robot approach to human beings and human behavior. At the time he wrote *Ghost,* there were few. Now there are many.

1.6.3.1
SELF

Hampden-Turner, Charles. *Maps of the Mind: Charts and Concepts of the Mind and Its Labyrinths.* New York: Collier Books, 1981.

Maps 40, 42, 45, 48, and 49 will familiarize the netweaver with cybernetic systems as applied to human psychology. In addition, this collection presents many other ways of mapping our insides (but certainly not *all* the ways).

Sayre, Kenneth M. *Consciousness: A Philosophic Study of Minds and Machines.* New York: Random House, 1969.

Sayre maintained, as many AI researchers still do, that an appropriate way to test and extend one's comprehension of a particular human mental function is to attempt to describe, in unequivocal terms, how that function might be implemented in a mechanical system. This book deals with the question of "machine consciousness" by pondering ordinary human consciousness.

Young, J. Z. *Programs of the Brain.* Oxford: Oxford University Press, 1978.

Step by step, Young guides the internal voyager through the programs of the biocomputer: controlling, coding, communicating, loving, caring, seeing, sleeping and dreaming, and consciousness. This book is filled with good introductory schematics of the brain, neurons, and brain systems. The discussion is broad enough for the nonexpert in brain research. Instead of providing detailed algorithms specifying how the brain works its wonders, Young discusses the broader features that are characteristic of biocomputer programs.

Lilly, John C. *Programming and Metaprogramming in the Human Biocomputer.* New York: Bantam Books, 1974.

Something of a cult book in its time. Lilly's model is downright useful in understanding and reprogramming self. It is made much more easily understandable now that microcomputers are becoming more accessible and widespread. This model leads directly into neurolinguistic programming (NLP) concepts.

1.6.3.2
GROUP

Reading about group dynamics is helpful only up to a point. After that, actual experience is the best teacher. People skilled in group leadership are rare, but virtually anyone can develop such skills. Effective managers, by and large, take an intuitive approach and seem, on the whole, to resist the idea that what they do in their various groups can be taught to others, much less be enhanced by reading some of the literature on the subject.

Borman, Ernest G., and Nancy C. Borman. *Effective Small Group Communication*. Minneapolis: Burgess Publishing Co., 1972.
This is the best practical handbook on the subject of group dynamics I have ever read. While there is nothing like experience in groups to teach the dynamics thereof, this slim introduction will give the netweaver a complete description of groups. It is clear and well organized.

1.6.3.3
ORGANIZATION

Gardner, John W. *Self Renewal*. New York: Harper & Row, 1965.
This and the books in the following section are all primary resources for those netweavers who go so far as to become interested in the development and renewal of organizations.

1.6.3.4
ORGANIZATION
DEVELOPMENT

Bennis, Warren G. *Organization Development: Its Nature, Origins and Prospects*. Reading, Mass.: Addison-Wesley, 1969.
This introduction to OD includes an overview of the educational strategies used. It is part of a series on OD.

Francis, Dave, and Mike Woodcock. *People at Work: A Practical Guide to Organizational Change*. La Jolla, Calif.: University Associates, 1975.
The authors identify eleven common "blockages" in organizations—things that prevent or slow down the work at hand—from poor recruitment to unfair rewards. They also present a "blockage questionnaire" that can help the netweaver identify the blocks in existing organizations or in new organizations that seem to be bogged down. Finally, after blockages have been identified, methods and educational techniques for clearing them are provided. Very practical. Highly recommended.

1.6.3.5
GLOBAL ACCESS
TO THE WORLD
GAME

See any of R. Buckminster Fuller's books—they all lead, eventually, to the World Game.

1.6.4
CYBERNETICS

Trask, Maurice. *The Story of Cybernetics*. London: Dutton, 1971.
A words-and-pictures book showing how cybernetics—ideas about communication and control—can be applied to every aspect of life. It is an especially useful introduction to this subject for creative visual artists.

Weiner, Norbert. *Cybernetics: or Control and Communication in the Animal and the Machine*. Cambridge, Mass.: M.I.T. Press, 1948.
A source text. The original. "One of the most influential books of the 20th Century."

Porter, Arthur. *Cybernetics Simplified.* New York: Barnes & Noble, 1970.

A nontechnical approach to the world of flowcharts, feedback, and servo-mechanisms. This one has a foreword by Marshall McLuhan. As you might expect, it's a good source for media buffs and those wishing to have a working knowledge of the field, but not a specialized one.

1.6.5
INFORMATION

Campbell, Jeremy. *Grammatical Man: Information, Entropy, Language and Life.* New York: Simon & Schuster, 1982.

In spite of its sexist title, if you read no other book on information theory, read this one. It is by far the best comprehensive synthesis of the subject presented in nonmathematical terms for the general reader. It is well written, succinct, and packed with stimulating possibilities for the imaginative who would go further.

Pierce, John R. *An Introduction to Information Theory: Symbols, Signals and Noise.* New York: Dover Publications, 1980.

If it weren't for information theory and theorists, we would not have color television or photographs of Jupiter. This book is both historically and technically valuable. Some math. An extensive glossary and an appendix on mathematical notation that should overcome any lack of such background on the part of the reader.

1.6.6
INFORMATION
SYSTEMS

Dertouzos, Michael L., and Joel Moses. *The Computer Age: A Twenty-Year View.* Cambridge, Mass.: M.I.T. Press, 1980.

Twenty experts contributed their sayings of sooth to this book. You think things are interesting now? Just wait.

Covvey, H. Dominic, and Neil Harding McAlister. *Computer Consciousness: Surviving the Automated '80s.* Reading, Mass.: Addison-Wesley, 1980.

If you haven't yet bothered to learn about microcomputers and how micro information systems are put together, this book should help. It is only one of many on the subject, but is full of good humor and sympathy for the newcomer. Netweavers may want to use this book in the education of clients and client systems.

1.6.7
NOÖSPHERICS

Teilhard de Chardin, Pierre. *The Phenomenon of Man.* New York: Harper & Row, 1959.

The work of a religious biologist who took Darwinism to a logical conclusion. Demanding but worthwhile. What is the Global Brain? Where is it? Is it just a metaphor and nothing more?

Stevens, L. Clark. *est: The Steersman Handbook: Charts of the Coming Decade of Conflict.* Santa Barbara, Calif.: Capricorn Press, 1970.

An extraordinary early (paleocybernetic) document that came hard on the heels of the Age of Flower Power. This book did more to liberate media-created hippies than just about any other work of its time. What is est? Electronic Social Transformation. Eco-Strategy-Tactics. Environment Systems Theory. Equilibrium of Sensory Thresholds. Earth Survival Techniques. And more.

1.7.3
THE INFOSPHERE

Youngblood, Gene. *Expanded Cinema*. New York: Dutton, 1970.
    Youngblood waxes turgid, in the manner of the mannerist aesthetician, in this early work. But it's all there, approached from a cinematic frame of reference: networks, the artist as designer, cybernetic cinema, and on and on. Many contemporary video and computer artists have been inspired by this introduction to noöspherics and the infosphere.

1.7.4
MEDIA

McLuhan, Marshall. *Understanding Media: The Extensions of Man*. New York: McGraw-Hill, 1964.

McLuhan, Marshall, with Quentin Fiore. *The Medium Is the Massage*. New York: Bantam Books, 1967.
    Though not without his subsequent critics and debunkers, before McLuhan few people thought much about media at all. Many of the ideas about media that have since become commonplace we owe to McLuhan. Money, clocks, and airplanes as media? Yes. And those looking for alternative names for videotext, teletext, and the rest are coining neologisms by the dozens: intermedia, immedia, micromedia, etceteramedia.

1.7.5
CONTENT AND
CONDUIT

Wirth, Timothy E., et al. *Telecommunications in Transition: The Status of Competition in the Telecommunications Industry*. A Report by the Subcommittee on Telecommunications, Consumer Protection, and Finance of the Committee on Energy and Commerce, U.S. House of Representatives (97th Cong.—Committee Print 97–V). Washington, D.C.: Government Printing Office, 1981.
    One of those government tomes that require selective reading, this report by the so-called Wirth Committee—the gadfly of AT&T—will give the netweaver an interesting slant on the telecommunications industry. If one must deal with the government, one must be armed with the appropriate information. This is a good source for such armament.

1.8
NEWS

Haigh, Robert W., George Gerbner, and Richard B. Byrne. *Communications in the Twenty-first Century*. New York: Wiley-Interscience, 1981.
    Will the news be affected by the new media? No doubt about it. Just how is debated. Conglomerate news organizations are jockeying for positions in the new Information Environment by adopting strategies that will position them where they think we'll be in years to come. The effect of the current media's handling of the "news" is to tell Americans *what* to think about and *how* to think about the what. This comes close to telling them what to think. This collection of essays brings together newspeople and information technologists to discuss not only the news, but the broader concerns of the age.

1.8.2
INTELLIGENCE

Dale, Arbie, and Leida Snow. *Twenty Minutes a Day to a More Powerful Intelligence*. Chicago: Playboy Press, 1976.
    Time management, self-motivation, meditation, information gathering, and communication: it's all sourced in this remarkably useful compendium. I've found this book to be constantly useful in reminding me of things that need to be balanced every day. Above all, human intelligence is a systems property, and

fine-tuning our subsystems gets us better system performance. No matter how many times before you've tried to "discipline" yourself, it's never too late to try new things. This book is a source of old things to try in new ways, bringing together techniques without ideological baggage.

**1.8.2.2
GOVERNMENT**

Burnham, David. *The Rise of the Computer State: The Threat to Our Freedoms, Our Ethics and Our Democratic Process.* New York: Random House, 1980.
There is a growing body of "computer nightmare" literature, to which this reference belongs. That is not to say that the fears are unfounded or that the writers of such books are crackpots. Orwell's primary instrument of repression was the two-way television set. As Walter Cronkite puts it in his foreword to this book, "Without the malign intent of any government system or would-be dictator, our privacy is being invaded, and more and more of the experiences which should be solely our own are finding their way into electronic files that the curious can scrutinize at the punch of a button." Netweavers should know these risks in order to help themselves—and their network associates—avert them.

Rowan, Ford. *Technospies.* New York: G. P. Putnam's Sons, 1978.
An earlier computer nightmare, in which our intrepid investigative reporter uncovers the "world of the technospies, the giant computer networks that collect and process information about each and every American from birth to death." Now the same computer networks (see also Jacques Vallee's work: [1.2.3]) are trying to convince us that "technospies" are KGB agents trying to infiltrate IBM. Or maybe technospies are a group of bright teens who are applying kitchen-table cybernetics (their home computers) to these networks. Such brouhaha should be seen by all netweavers for what it is: a diversionary tactic. Keep your eye on those who are raising the most ruckus about "computer crime." Many of them are committing far greater crimes in the name of ideologies cloaked as "national security" or even "computer security." If we are to place "legislative controls" on computers, let such "controls" *begin* with the government and corporate network levels, where the abuses and potentials for abuse are greatest.

**1.8.2.3
MILITARY**

Sargant, William. *Battle for the Mind.* New York: Doubleday, 1957.
What do psychologists, brainwashing, cults, politicians, and military indoctrination have in common? This is almost a handbook for would-be mind-manipulators. The seamier side of cybernetics as applied to individuals and groups.

**1.8.2.4
MACHINE**

Taube, Mortimer. *Computers and Common Sense.* New York: McGraw-Hill, 1961.
Dr. Taube's arguments are still being used against the very *concept* of AI research, which he classes as "scientific aberration." Machines never have thought. Still don't and never will. Certainly. And they don't gather intelligence. Right.

**1.9
NETWORKS**

Re: Classification of Networks. The exploding universe of options and possibilities, combinations and permutations of information, hardware, software, and networks makes it difficult to organize and classify information as such info is generated. The scheme used in this book has been developed based on principles from General Systems Theory. Even so, some things might not be neatly "captured" by this scheme. This is especially true in any attempt to classify networks. Some, for example, make distinctions between local area networks, adding a level below that called "small area networks" and "micro area networks." As the information environment convolutes, takes on strange hybrid shapes, inventors and manufacturers alike invent new categories to differentiate themselves and their products from others. This is called "creating a market niche" or "positioning."

Still others classify networks more according to sociological ideal types, differentiating only two kinds of network: "mass" and "vernacular." The network classification used here assumes rough divisions according to the relative size of a given network in both geography or "reach" and in number of users, as well as the size of the information systems that are its constituent parts.

The Netweaver's Sourcebook concentrates on individual and business aspects of the information economy, hence we only bow briefly at the altars of broadcast media networks. Besides, the metaprinciples presented herein apply to all sizes and shapes of networks, no matter how they are classified.

1.9.2 NETWORKING

Welch, Mary-Scott. *Networking: The Great New Way for Women to Get Ahead.* New York: Warner Books, 1980.

This tome on the popular conception of networking—a slightly more sophisticated form of power building and social climbing—can be applied by anyone, not just women, wishing to get ahead. Ultimately, this kind of networking is applied within existing organizations, usually in corporate environments, and leads to the formation of a group of some kind. The links are purely face-to-face and telephone, but many of the principles can be applied to micro-based networks and computer conference organizing.

1.9.3 NETWEAVING

James, David. "Networking: A Powerful Tool for Personal Communication." *Personal Computing,* January 1983, p. 46.

This is really about net*weaving,* not net*working.* The article brings the personal and the technical together in an overview.

1.10 SYNCHRONICITY AND GENERAL SYSTEMANTICS

Jung, C. G. *Synchronicity: An Acausal Connecting Principle.* Princeton: Princeton University Press, 1960.

This is an essential source document on the subject.

Malaclypse the Younger. *Principia Discordia.* Mason, Miss.: Loompanics Press, 1979.

What is the relationship of chaos to order? Uneasy. Are we governed by chance or necessity or both or neither or sometimes? The answers are all here, my friend. Hail, Eris!

Personal Systems

"It's all well and good to know about myself," the skeptical netweaver maintains, "but right now I'm interested in *technology,* not all this namby-pamby self-development rigmarole." If you are already somewhat familiar with microcomputing, fiber optics, lasers, genetic engineering, space exploration and satellites, fusion energy and the brain sciences, then go immediately to Chapter 4 of this source-book and start soaking up more technical specifics. On the other hand, if you need to do a refresher walk through our "high" tech age, turn to the following.

Dowling, Colette, and the Print Project. *The Techno/Peasant Survival Manual: The Book That De-mystifies the Technology of the '80s.* New York: Bantam Books, 1980.

On the other hand, you may not like *any* of the resources (I seriously doubt it) included in this appendix. If that's the case, then I recommend the next title.

Popenoe, Cris. *Inner Development: The Yes! Bookshop Guide.* Washington, D.C.: Yes! Inc., 1979.
    Popenoe's guide is nothing more or less than a compendium of reviews of more than 10,000 books from many publishers throughout the world. One can learn from it more than just where to get what books on self-knowledge and self-development. Popenoe is about to do the same thing for computer books. The guide is available from:

Yes! Inc.
1035 31st Street, N.W.
Washington, DC 20007

Many of the books listed in the Yes! guide can be borrowed by mail or ordered directly from:

The Lucis Trust Library
866 United Nations Plaza, Suite 566—7
New York, NY 10017

Then, in order to maintain an overview on the cutting edge of the brain sciences, a subscription to:

Brain/Mind Bulletin
P.O. Box 42211
Los Angeles, CA 90042

Brain/Mind editor Marilyn Ferguson built this excellent newsletter on her network of contacts in the field. While *Brain/Mind* is thorough and useful, the same cannot be said for its companion publication, *Leading Edge,* which generally isn't.

2.1.2
BRAIN CENTERS

Ryan, Paul. *Cybernetics of the Sacred*. New York: Doubleday Anchor, 1974.

How is the six o'clock news format a means of social control? Why is this format being adopted by emerging television networks around the world? Our media don't have to be used this way. Ryan used videotape to arrive at some remarkable conclusions about human contact through communication. The insights he shares are easily generalizable to the newer micro communications systems.

Brain centers in the form of connected networks of people working toward similar planetary, regional, and local goals are listed in the directory of the Unity-in-Diversity Council. Get more information from:

Unity-in-Diversity Council
626 Echo Park Ave.
Los Angeles, CA 90026

2.1.3
INPUTS

There are hundreds of computer magazines now. Scanning the newsstands and only buying those issues that have information in them that you particularly want is probably the best way to "keep up" with the literature. In addition, there are several good periodical guides specifically dedicated to computer and tele-communications publications. Trade magazines are wonderful sources because many of them are free if you fill out a "qualifying" card or questionnaire.

Many journals and newsletters are high-priced not because information in them is not available anywhere else, but because their "target market" is in the Fortune 500 class, and those people are still held sway by the "get what you pay for" stereotype: if it doesn't cost at least three figures, it isn't worth the executive's time. It doesn't take much today, given the low cost of micro information systems, to start a newsletter on *any* topic. Price it attractively, and the chances are it will be successful, especially if it is an alternative to another, more expensive source in the same—or similar enough—information domain.

2.1.4
FILTERS

Koberg, Don, and Jim Bagnall. *Values Tech*. Los Altos, Calif.: William Kaufmann, 1976.

This guide can assist the budding and seasoned netweaver alike in discovering what is valued and valuing what is discovered. Practical success in any endeavor involves not only a pragmatic approach to problem solving, but also having a knowledge of values and decisionmaking theory. This open-ended course can provide such knowledge, when vigorously self-applied.

**2.1.5
SELF-COMMUNI-
CATION AND
CREATIVITY**

Scheflen, Albert E. *How Behavior Means: Exploring the Contents of Speech and Meaning—Kinesics, Posture, Interaction, Setting and Culture.* New York: Jason Aronson, 1973.

Do you wave your hands when you talk? Do your eyebrows send signals you don't know about? How're your skinetics? If you had to go on television tomorrow to explain yourself, could you get a date? Scheflen provides a complete view of the systems of communication used by our biocomputers. Personality and individuality play smaller roles in communication than we would perhaps like to think. Of equal or greater importance are the cultural and life-context determinants of our communications.

Simon, Sidney B., Leland W. Howe, and Howard Kirschenbaum. *Values Clarification: A Handbook of Practical Strategies for Teachers and Students.* New York: Hart Publishing Co., 1972.

This resource was designed for use in the classroom, but it is widely applicable in other settings as well. The mere activity of *examining values* consciously is often enough to change them, a fact that makes conservatives of all stripes somewhat opposed to the use of values clarification techniques in public schools. *Values Clarification* presents methods and practices, not particular values. It is grounded in practical experiences rather than philosophy, one way or another.

**2.1.6
MEMORY AND
PROSTHESES**

If you are one of the million or so people who have diabetes, or if you have allergies or other conditions that may be important to know about you in a medical emergency, then there is already an information prosthesis you can carry around with you. It's called the Medical Information Card. The card is about the size of an ID or credit card, so it fits neatly in your wallet or purse. The card contains your medical history reproduced on microfilm. Emergency personnel across the country are being alerted to look for special stickers on wallets, windshields, and backpacks that will tell them you're carrying the card. The card has its own built-in reading lens. This is a magnificent way to use information technology. Send a stamped, self-addressed envelope for more information to:

National Health and Safety Awareness Center
Application Dept.
333 N. Michigan Avenue
Chicago, IL 60601

**2.2
PROCESSING:
PROGRAMS AND
METAPROGRAMS**

Bateson, Gregory. *Steps to an Ecology of Mind.* San Francisco: Chandler Publishing Co., 1972.

Nowhere has the influence of cybernetics and information theory been more apparent in the humanities than in this classic work of Bateson's. Understanding patterns and relationships is the essence of systems science. It is also useful in any endeavor that seeks to create new patterns and relationships in the Information Age. These steps will bring together your current understanding in new ways. If you have already read Bateson's seminal book, I encourage a reread in parallel with your reading of this *Netweaver's Sourcebook.*

2.2.1
LIFETIME
LEARNING:
PROCESSES
AND CHECKLIST

Anderson, Marianne S., and Louis M. Savary. *Passages: A Guide for Pilgrims of the Mind.* New York: Harper & Row, 1973.
   A delightful and beautifully designed developmental program. Skills in habit breaking and habit making. Skills in interpersonal relationships. Skills in creativity and greater awareness. Marvelous!

Gross, Ronald. *The Independent Scholar's Handbook: How to Turn Your Interest in Any Subject into Expertise.* Reading, Mass.: Addison-Wesley, 1982.
   It is possible, in fact now more than ever, to achieve expertise (in the positive sense of the term) *and* recognition while working outside the university, government or the corporations. Ron Gross helps give direction, avuncular support and just plain good advice for anyone who wants to apply rigor to his or her independent scholarship or research.
   Rather than waiting for schools to go on-line and try to sell their services and resources to you, use your netweaving tools to build your own scholarship network, making the system an ongoing part of your lifetime education. Start, or continue, with this *Independent Scholar's Handbook.*

2.2.2
GOAL SETTING,
GOAL SEEKING

Many people find that beginning the process of finding meaningful work, changing a career, or inventing a new career is at the head of the list of personal changes they desire. The processes employed in finding employment (right livelihood, if you will) are identical in form to those involved in any significant self-change. Goal setting, in this case, involves economic and values objectives.

Bolles, Richard Nelson. *What Color Is Your Parachute? A Practical Manual for Job-Hunters & Career Changers.* Berkeley, Calif.: Ten Speed Press, 1976.
   I know people who have invented new jobs for themselves based on Bolles's methods. I know others who have found new niches in the corporate world based, also, on Bolles's recommendations. I know others who ignored Bolles in seeking gainful employment. Oh, they eventually got jobs, all right. But were they happy? And how long did it take them? Not too. And a long time, indeed.

   Existing telecommunications network-based services can also be used to find new jobs. Two representative ones are Connexions and CLEO.

Connexions
55 Wheeler Street
Cambridge, MA 02138
800-562-3282 (National)
800-JOB-DATA (Customer Service)
617-497-4144 (Mass.)

Connexions was started by a couple of entrepreneurial execs who used to be employed by the likes of Honeywell and Arthur D. Little. The service is available through the Telenet packet network from personal micros. The charges for the job seeker are nominal ($15 for two cumulative hours on the system as of this writing) and give you entrée for up to a year. You can read, write, and upload your résumé, and transmit it privately to firms of your choice.

Client firms are mostly high-tech companies based in the Boston area. Connexions also has a branch office in Cupertino, California (home of Apple Computer, Inc.).

To connect with Connexions, dial your local Telenet number, hit two RE-TURNs when you get the carrier signal, then enter your terminal type or just another RETURN. The access code you'll be asked for is C 60366, and the user ID number of JOB12345. For human contact, or to make sure these instructions are still correct at the time of publication, call 617-497-4144.

For CLEO (Computerized Listings of Employment Opportunities), the micro access numbers (300-Baud, full-duplex ASCII) are:

415-482-1550
408-294-2000
818-618-8800
714-476-8800
619-224-8800

For access assistance: 818-618-1525

CLEO is a service of Copley Press, Inc. The bias is heavily California, with jobs in engineering and data processing prevailing. It doesn't charge job seekers, but client companies, who place their ads in the CLEO system.

There is room in most regions of the country for more of these kinds of services, including more personalized ones based on micro networking systems.

Watzlawick, Paul. *The Language of Change.* New York: Basic Books, 1978.

How do you talk to yourself? Do you send positive messages to yourself, or are you constantly programming fear, anxiety, and apprehension? This book can help you discover how *your* maps and models—especially the unconscious ones—are put together. You then have a better shot at changing what you want changed.

2.2.3
PLANNING:
PLANNING AIDS

Fabun, Don. *The Dynamics of Change.* Englewood Cliffs, N.J.: Prentice-Hall, 1967.

Many of the trends popularized by writers such as Toffler *(The Third Wave)*, Naisbitt *(Megatrends)*, and the Club of Rome were apparent to reflective thinkers as early as the 1930s and '40s. Aldous Huxley and others, for example, wrote about the population and energy systems crises long before they became popular political issues. Fabun's book is another precursor of the futurist books of today. He explores the outlines and parameters of change itself, using concrete examples in topics ranging from "telemobility" (telecommuting) and automation, to leisure and the unforeseen consequences of any change. This reference is a wonder in itself, since Fabun used all his skills as a graphics designer in bringing it to print.

Whittlesey, Marietta. *Freelance Forever: Successful Self-Employment.* New York: Avon Books, 1982.

When you are self-employed at anything, you are your own employer, employee, and product. The sections on controlling time, organizing your work (soft) space, and budgeting your unpredictable income are particularly helpful for would-be professional netweavers. An example of advice I wish I had had when I first went freelance: "Discipline yourself, but realize that there will be fluctuations in your level of discipline. Don't expect such perfect adherence to your own rules that a lost day causes you to feel you've failed completely. Creative minds work in special ways. Occasionally you need a 'lost' day. Accept this, and then get back to work the next day."

2.2.4
HUMAN LIFE-STYLING: DESIGN

McCamy, John C., M.D., and James Presley. *Human Life Styling.* New York: Harper & Row, 1975.

True health care, maintain these authors, aims at the prevention of future diseases, not their treatment after they've developed. In addition, they make the essential connections you need to relate the environment (the apparently not-you) to your health care and management. You can do something about both working alone and with others. From the environmental hazards to nutrition and stress, this guide is a great starting point for the new and better you.

Kirn, Arthur G. *Lifework Planning.* Hartford, Conn.: Arthur G. Kirn & Associates, 1974.

One of the most valuable and productive of the self-change systems I have used, *Lifework Planning* is thorough yet gentle on the mind. You get to look at where you are, where you've been, and where you're going. Who needs you? What are your values? What means something to you? All the exercises are experiential and interactive and real-time. This means that you'll have to do *work* to get anything out of these methods. But what rewarding work it is!

2.2.5
TIME MANAGEMENT

There are literally hundreds of time management books, pamphlets, and consultants. The task of finding the one(s) that work for you has to be integrated with your self-design process. Most systems will require some change in your consciousness of time itself and therefore in how you treat "it." Keep a calendar. Stick to it. That's it.

2.2.6
CREATIVE LEISURE AND STRESS MANAGEMENT

Osborn, Alex F. *Applied Imagination: Principles and Procedures of Creative Thinking.* New York: Charles Scribner's Sons, 1957.

It may be that things like imagination and creativity cannot be taught. You either have it or you don't. You're either a Great Artist or you're not an artist at all. On the other hand, you may believe wholeheartedly in the universality—the democracy—of the imaginative faculty. If the latter, you may follow up on your beliefs by running out and buying up a library full of books on how to be creative. This early work is comprehensive. It's all here. What has followed in the field has been mostly repackaging and refinements on these basic ideas. And even if imagination is the sole province of the gifted few, these procedures certainly won't *hurt* anyone who tries them.

Parnes, Sidney J., Ruth B. Noller, and Angelo M. Biondi. *Guide to Creative Action: Revised Edition of Creative Behavior Guidebook.* New York: Charles Scribner's Sons, 1977.

A master of *science* degree in "Creative Studies"? Certainly. This particular guidebook is another rich resource in the study of self-actualization and creativity. Creative behavior *can* be deliberately cultivated. No long-term studies have been done comparing those who claim to do so and those who don't.

Beware of dogmatic "schools of thinking" that have repackaged creativity under their own definition of what "thinking" is. Most of these schools will have you "thinking by the book," and that's not what creative problem solving is all about at all.

**2.3.2
BIOFEEDBACK
AND
SUGGESTOLOGY**

Work on enhancing learning is going on in the U.S. and elsewhere. Find out more from·

SALT (Suggestive-Accelerated Learning and Teaching)
Box 927
Iowa City, IA 52240

**2.3.3
STATES OF
CONSCIOUSNESS**

Schwarz, Jack. *Voluntary Controls: Exercises for Creative Meditation and for Activating the Potential of the Chakras.* New York: E. P. Dutton, 1978.

This man has gained remarkable mastery over his own nervous system. He shares some of his models and methods in *Voluntary Controls.* He monitors his energy and behaviors carefully, and uses the methods he preaches. However, unlike many teachers, he encourages you to experiment for and with yourself, which may include discarding his recommended exercises and designing your own.

Carrington, Patricia. *Freedom in Meditation.* New York: Anchor Press/Doubleday, 1977.

After the early psychedelic wave of interest in things Eastern (gurus, meditation, mantras, and the rest), the scientific and medical communities began taking an active interest in the roles that mental states play in physical health. Carrington provides the "self-permission" needed to embark on one's own journey of internal exploration without dogmatically induced guilt. Regular self-communion through some form of meditation is one of the most powerful self-change agents a netweaver can employ. The relationship between meditation, relaxation, desensitization, and self-education is clearly set forth here.

**2.3.4
COMMUNITIES OF
CONTELLIGENCE**

Watzlawick, Paul, John Weakland, and Richard Fisch. *Change: Principles of Problem Formation and Problem Resolution.* New York: W.W. Norton, 1974.

Work in many communities of special interest will involve "solving a problem" of one kind or another. Sometimes the apparently "commonsense" or "logical" approach to problems don't work. Sometimes a lot of money and human energy are wasted on "solutions" that do little for anyone other than those who got the grant money. Change is the name of the problem game. Sometimes it occurs spontaneously—apparently without outside intervention. The work presented in this book goes far beyond the interpersonal and has national and international applications as well. I would say that this book is absolutely required for any netweaver who wants to promote a new network or make contributions to old ones.

Bear, John. *Bear's Guide to Non-Traditional College Degrees*. Oakland, Calif.: Rafton & Bear, 1980.

Let's say you've turned yourself into an expert in some information domain. You've done most of your learning on your own, in the "real world" of application and experience. Getting acknowledgment for your accomplishments may still require that you get a degree. Dr. Bear's guide can help steer you to a self-designed degree program from a variety of institutions in many states. Contact him at:

Dr. John Bear
P.O. Box 11447-RH1
Marina del Ray, CA 90295

Does your life-style fall into a category? SRI International, a brain center in Menlo Park, California (94025) thinks so. They've identified (created a typology for) nine major life-styles in four categories:

- Need-Driven
  Survivor life-style
  Sustainer life-style
- Outer-Directed
  Belonger life-style
  Emulator life-style
  Achiever life-style

- Inner-Directed
  I-Am-Me life-style
  Experiential life-style
  Socially conscious life-style
- Combined Outer- and Inner-Directed
  Integrated life-style

SRI is doing research in these VALs (short for values, attitudes, and life-styles) using findings from developmental psychology and sociology and market research. The typology, in fact, is most useful to marketeers within corporations looking to find new ways to push people's "hot buttons" through advertising. If you key into a person's basic values, you are more likely to be able to sell something to him.

The VALs life-style types are one way to identify communities of contelligence. There are, of course, many other ways. A pointer network of such communities puts out a quarterly journal called *In Context*. This "quarterly of human sustainable culture" is full of integrating ideas from ecology, philosophy, psychology, technology, and politics. More info from:

In Context
P.O. Box 30782
Seattle, WA 98103

**2.4
TERRITORY AND
INFORMATION**

Bakker, Cornelis B., and Marianne K. Bakker-Rabdau. *No Trespassing! Explorations in Human Territoriality*. Corte Madera, Calif.: Chandler & Sharp Publishers, 1977.

A better understanding of your own territorial propensities can help you in getting along with spouse, roommates, business associates, and friends. You will also gain a better grasp of how and when to use criticism, lament, protest, and complaint. Learn how to expand or give away your territory and still keep your wits. Do you cope well with generosity? Can you overcome envy and jealousy? As you might expect, how we manage our territories depends on how we use and control our communications.

Garvin, Andrew, and Hubert Bermont. *How to Win with Information or Lose without It.* Washington, D.C.: Bermont Books, 1980.

Don't let its sappy title put you off. This is nothing like the me-decade "How to Win" or "Winning through . . ." fad books. The authors teach that the most important part of an information system is *you.* (What have I been saying, huh?) Their "generalist's source list" includes only those likely to be of interest to businesses.

Once you have identified your information territory and have begun to create new information within it, how do you get support and benefit from what you have done? It's no fun being a Starving Artist, and in the Info Age there's no reason to be. Protecting your intellectual assets is as important as protecting your financial assets. The former, in fact, are part of the latter. If you are a consultant, one of your more important assets is your network of contacts who provide you with job leads or the jobs themselves. You can't copyright your mailing list, of course, but you can at least treat it with the protection it deserves.

If you work in existing media, as a writer, producer, composer, or programmer, or in some combination of these Info Age skills, you'll need to know about copyright, how the new technologies are affecting your legal rights, how to protect yourself in the contract negotiation phase of your work, and so on. The following references are all pointer documents to more info resources and the ins and outs of the particular media involved.

Bronfeld, Stewart. *Writing for Film and Television.* Englewood Cliffs, N.J.: Prentice-Hall, 1981.

Bunnin, Brad, and Peter Beren. *Author Law and Strategies.* Berkeley, Calif.: Nolo Press, 1983.

Remer, Daniel. *Legal Care for Your Software: A Step-by-Step Guide for Computer Software Writers.* Berkeley, Calif.: Nolo Press, 1982.

**2.4.1
PRIVACY**

There is more on privacy in Chapter 3.0 and Appendix III. From a personal point of view, privacy is something that each of us has to defend in our own way(s). Becoming aware of the ways in which our privacy is eroded by new technologies is a good first step. There are networks devoted to the issues, and the issues will likely arise on your network(s). A good newsletter reporting on things like the abuse of mailing lists, lie detectors, and public records is:

Privacy Journal
P.O. Box 8844
Washington, DC 20003

**2.4.2
ETHICS AND
ETIQUETTE**

We are fast becoming a nation of "little brothers" in which everybody who can afford to hire an ex–government agent can spy on anyone or anything else, including government itself. All the laws in the world won't replace respect for the privacy and communications rights of others in our personal spheres. New technologies mean new etiquettes.

**2.5
PERSONAL
SOFTSPACES**

Bolt, Richard A. *Spatial Data Management.* Boston: M.I.T./Architecture Machine Group, 1979.

This little volume is packed with ideas for netweavers. Based on work sponsored by the defense establishment, *Spatial Data Management* shows how sound, light, and touch can be integrated into an information system that is highly personal and subjective. The basic metaphor for organizing information in this system is "space." You access data items by going where they are rather than referencing them by name. In a sense, this is how the entire information environment is becoming structured. The desktops and windows showing up on computer screens got a big push from this book. Although it was published in a limited edition, you may still be able to obtain a copy from M.I.T. directly or find one at your local university library.

**2.5.1
HARDSTUFF**

For more specifics on micro communications hardware, see Chapter 4.0 and Appendix IV. Also, just about every consumer-oriented micro magazine runs frequent "how to select your hardware" articles, complete with comparison charts. I recommend that you prepare your own comparison charts based on the planning and analysis you do on your personal needs and goals. Of course, it never hurts to ransack the work of others, but anything done by someone else can obscure your own choices and options.

**2.5.2
SOFTSTUFF**

See Chapter 4.0 and Appendix IV.

**2.5.3
ERGONOMICS**

Knapp, John M. "The Ergonomic Millenium." *Computer Graphics World,* 6:86, 1983.

This article cites the NIOSH studies that found 93 percent of CRT operators reported eye strain; 90 percent reported neck pain; 89 percent had headaches; 83 percent had "severe fatigue"; 79 percent had "burning eyes"; and 72 percent reported blurred vision. Many combinations of poor lighting, poor seating, and poor human relations in the work environment can produce these symptoms. Additionally, what is "good" for one person, of a given physical size and psychological makeup, may be entirely debilitating for another.

For an ongoing view of what designers are doing in the way of good and bad human factors engineering, there is a newsletter devoted to ergonomics:

The Ergonomics Newsletter
The Koffler Group
3029 Wilshire Blvd.
Suite 200
Santa Monica, CA 90403
818-453-1844

Sommer, Robert. *Personal Space: The Behavioral Basis of Design*. Englewood Cliffs, N.J.: Prentice-Hall, 1969.

This early reference works at the juncture of architecture and psychology. Our sense of personal space, private space, intimate space, working space, and public space can have a big impact on how we feel and behave. Personal spaces and territory must work together.

If you become extensively involved in designing softspaces, including things like desks and terminals and chairs, you'll want to pay more attention to the physical needs of your specific clients, including children and the physically challenged. *Humanscale* is the best and most succinct set of design tools you can have. It is available from:

Humanscale
M.I.T. Press
28 Ca iton Street
Cambridge, MA 12142

While in the final stages of preparing the manuscript for this book, the author ignored all this ergonomic advice and began spending more hours at the video screen than was wise. He suffered.

2.7 IMAGINING INFORMAGI- NATION

DeMille, Richard. *Put Your Mother on the Ceiling: Children's Imagination Games*. New York: Viking Press, 1973.

The exercises in this slim, imaginative book encourage children and childlike adults to prolong and preserve their penchant for taking liberties with reality. Get in touch with your most neglected yet most enjoyable parts.

Maue, Kenneth. *Water in the Lake: Real Events for the Imagination*. New York: Harper Colophon Books, 1979.

Open the book to any page. Read as much or as little as you like. Skip around from place to place. Continue. Get thirteen walnuts. Place them in your environment. Take great care to place the walnuts well. Perform thirteen gestures of good fit. With the walnuts. Without the walnuts.

There is much more here than meets the I with print.

2.8 RESULTS: THE PRAGMATICS OF COMMUNICA- TION

Watzlawick, Paul, Janet Helmick Beavin, and Don D. Jackson. *Pragmatics of Human Communication*. New York: W. W. Norton, 1967.

The effects of a communication are the behavior(s) it elicits. This early work, focusing as it does on the paradoxical and pathological, heightens one's awareness of the importance of interactions between human beings. In the months and years to come, computer-mediated communications will be just one *mode* in a mix of message transfer systems between people. This presentation is worth reading if only for its epilogue, in which the authors suggest that "life is a partner whom we accept or reject, and by whom we feel ourselves accepted or rejected, supported or betrayed."

Aranguren, Jose Luis. *Human Communication*. New York: McGraw-Hill, 1967.

Communication = transmitter, transmission/medium, message, receiver/decoder. Anthropology, language arts, applied sciences, secular and religious rites, information, art, sculpture, music. A grammar of modern communications has to include all of these, as well as punched cards and floppy disks, village gossip and telecommunications. Aranguren believes that our Information Age society will be one of "continuous education," using, but not used by, the new sciences of automation, information and prevision (futurism). *Human Communication* is itself a masterful piece of communication backing up the author's optimism.

Art and information technology are doing interesting things to each other. Some of what's happening among those who consider themselves "dedicated to the arts" is being reported by *Art Com,* a periodical covering video, performance art, cybernetic art, and new media syntheses. Available from:

Contemporary Arts Press
LaMamelle, Inc.
P.O. Box 3123
Rincon Annex
San Francisco, CA 94119

APPENDIX III

Social Systems

Wiener, Norbert. *The Human Use of Human Beings: Cybernetics and Society.* New York: Doubleday, 1950.

This is still a seminal sourcework on the relationship(s) between our social macrosystems, communications, and information theory.

**3.2.1
INFORMATION
ECONOMICS**

Libraries as sources and repositories of information are not, by any means, becoming obsolete. The problem is, now, how to finance "going on-line" for libraries. This has two faces: the searching face, in which libraries use computers to search other libraries and info sources for its users, and the providing face, which enables libraries to, for example, put their card catalogs on-line for access through micros. It is important, as we enter an age of widespread use of the telecommunications nets to get information, that we preserve the tradition of free public access to information that libraries have upheld. There is still an awful lot of information that will be relegated to books for some time to come.

There are two networks of library-related information sources. ALANET, from the American Library Association, provides e-mail and information services for ALA members. ALANET's services are provided through ITT Dialcom, of Silver Springs, Maryland. The system is Telenet, TYMNET, telex, and Omninet accessible. For sign-up information, contact:

Alanet
ALA Headquarters Library
50 E. Huron Street
Chicago, IL 60611

The second service, called CLASS (Cooperative Library Agency for Systems and Services), provides info concerning micros and telecommunications to libraries.

CLASS
1415 Koll Circle, Suite 101
San Jose, CA 95122
408-289-1756

3.2.2
INFORMATION
TECHNOLOGY

Thompson, Gordon B. "Memo From Mercury: Information Technology *Is* Different." Montreal: Institute for Research on Public Policy. Paper No. 10, June 1979.

The problem of defining a new economic system for the Information Age is the major problem of our time, and it is being approached haphazardly and in a non-systems-oriented way. This is ironic, to say the least. Thompson's early paper is typical of the approach so far: taking the qualities and principles of information and information theory and attempting to extrapolate to what we think *may* be the case during and after the current economic transformation. Alas, I suspect that what we must do is create a consensus concerning *how it must be* and *how we want it to be*. That is, we must bring together all the segments of our society, all our special interests, and, literally, define and then create a *new economic system,* based on what we are able to do with information technology. This does not mean revolution, socialism, or capitalism. I suspect that those terms must be jettisoned in favor of a more pragmatic approach. *We have to redesign the macrosystem we call by the name economics.* This redesign would use all the systems and design and creativity techniques our culture has documented so far. The design process would have to be ongoing. In all fairness, Thompson and others have probably done more up-to-date work on the information economy that hasn't yet breached my own softspace filters.

Rosenthal, Lois. *Partnering.* Cincinnati, Ohio: Writer's Digest Books, 1983.

The next step in organization, after working independently and for yourself, is the partnership. Partnerships can be dyads—two people—or more, up to legal limits set by the states. Partnerships aren't only for businesses. Sailboats and homes can be owned by partners, too. *Partnering* maps out a course to successful partnerships that won't break down at the first sign of crisis or conflict. The book includes self-administered questionnaires designed to help you and your prospective partner(s) determine whether you're compatible. Sample written agreements and case histories are included.

Bowers, David G. *Systems of Organization: Management of the Human Resource.* Ann Arbor: University of Michigan Press, 1976.

So your little infocompany has grown, gone national, and you're ready for the next stage of business growth: adding more people. This reference will help you develop a participative management style and a noncoercive organization from the git-go.

Support for the would-be entrepreneur and small businessperson can be obtained from a variety of sources. Entrepreneur's Alliance, in California, is a prototype social network (not yet micro-accessible) of those starting up new enterprises and those dedicated to supporting such new efforts, including venture capitalists and consultants of various stripes. This kind of support will be increasingly useful in just about every community in the country as the Information Age proceeds. Local and national networks of self-starters can help create the 20 million new jobs we need just to stay "voluntarily simple," much less prosperous. For more:

Entrepreneur's Alliance
1333 Lawrence Expressway
Suite 150
Santa Clara, CA 95051
408-246-1007

**3.3.2
SYSTEMS
ANALYSIS**

Thierauf, Robert J. *Systems Analysis and Design of Real-time Management Information Systems.* Englewood Cliffs, N.J.: Prentice-Hall, 1975.

This is a bible for those working in a corporate environment in MIS (Management Information Systems) or DP (Data Processing) departments. It takes you from feasibility study through implementation of systems for handling accounting, personnel, manufacturing, inventorying, purchasing, and so on. Mathematical models are supplied throughout.

**3.3.3
SYSTEMS DESIGN**

No single directory is "stand-alone." If you are going to be designing systems based on microcomputers, you'd do well to have several different kinds of directory on hand: software, network services, hardware, and so on. For its price, the *Bowker/Bantam 1984 Complete Sourcebook of Personal Computing* (New York: Bantam Books) will be a good companion to the book you are holding in your hands, in spite of the fact that the word "complete" in its title is suspicious at best, grandiose at worst.

**3.3.4
PRIVACY**

Do you get junk mail you don't want? This minor invasion can be overcome. Simply write to:

Direct Mail Marketing Association
6 E. 43rd Street
New York, NY 10017

Tell them you want your name taken off all mailing lists. They'll distribute your name to list brokers who will comply. They probably won't get your name off *all* the lists that everyone is keeping, but it'll go a long way.

On the other hand, if you want to *get* information about certain types of products or services, the same organization will see to it that your name is put on such lists. Junk mail, properly filtered, can be a great source of essentially *free* information for your business.

Using mailing lists to promote yourself, a product, or a service, or to make direct mail order sales, is a long-standing American business tradition. List suppliers register with the DMMA, and you can find out who these suppliers are and what kinds of lists they have through the DMMA's *List Supplier Register.* Write them for details.

**3.3.5
LIBEL AND
SLANDER**

Sanford, Bruce W. *Synopsis of the Law of Libel and the Right of Privacy.* New York: World Almanac Publications, 1977.

The nice thing about being a writer, reporter, or commentator in the United States is that you can speak your mind. Then you get sued. The mark of the amateur writer (especially those who get their hands on their own newsletter or magazine column) is the tendency to use his or her newly found power of the pen to lash out at people, places, and things that annoy them. That phase usually passes, but it's the period during which the new writer may be most vulnerable to a lawsuit on the basis of invasion of privacy or libel. This booklet can help keep you out of trouble, yet still allow you to speak your mind about behemoth egos and other nuisances. People who participate in computer conferences, community electronic bulletin boards, and similar computer-based media need to know about these laws and their responsibilities when it comes to messages about other people. System operators (sysops) also need to have this information. It is also covered in *Author Law and Strategies* [Appendix II: 2.4].

3.4
GROUPS

Prince, George M. *The Practice of Creativity.* New York: Collier Books, 1970.
The approach is called *synectics,* and it provides a means for bringing problems and creative people together in ways that have real-time results. Many of the methods and techniques, insights and overviews concerning creativity and group problem solving contained in this early reference can be found elsewhere, of course. But this little handbook is a nice alternative to *Robert's Rules* for running meetings.

Bureau of Business Practice, Inc. *Quality Circles: A Dynamic Approach to Productivity Improvement.* Waterford, Conn.: Bureau of Business Practice, 1981.
The quality circle approach in manufacturing has a history going back to the days of the National Training Laboratories in Bethel, Maine. The Japanese, after World War II, ever in search of better techniques to adopt into their culture, imported the quality circle. Then, as they have done with tangible goods, they exported their tested and improved version. It is this version that has been adopted and adapted by U.S. companies such as J. C. Penney and Unitek (a dental supplies manufacturer) and in the aerospace industry. This booklet, for all its arid prose, presents the essentials of the QC in a pamphlet-length sixty-four pages. If you are faced with having to obtain top management support for a QC—a small network—program within a company, this is a safe intro and won't intimidate the nonreader, whether manager or union leader.

Halprin, Lawrence, and Jim Burns. *Taking Part: A Workshop Approach to Collective Creativity.* Cambridge, Mass.: M.I.T. Press, 1974.
Workshops that are well done are pleasures indeed, like well-performed symphonies or operas. Entertainment, education, socializing, and creative action are all "products" of good workshops or seminars. This delightful reference may be the only one you need to help create such events. The authors brought their experience of many years' worth of workshops to bear on *Taking Part.* Give this book to the organizer(s) of the next boring workshop you are tricked into attending.

3.4.2
DECISIONMAKING

Many of our tasks in groups involve a division of labor. Each member of the group has some expertise and is responsible for a specific segment of the *task territory.* It is important that each member of the team have a specific understanding concerning what his or her decisionmaking territory will be. It can become highly unwieldy for any group to try to make *all* its decisions by consensus. Generally policy decisions will be made by consensus and individuals will then have territorial responsibilities in carrying out the policies. For more on this aspect of group decisionmaking, see *No Trespassing!* [Appendix II: 2.4].

3.5
THE BIZ BIZ

Lesko, Matthew. *Information U.S.A.: The Ultimate Guide to the Largest Source of Information on Earth.* New York: Penguin Books, 1983.

Every scholar, expert, and citizen ought to have a copy of this book in his or her personal information data banks. According to Lesko, our best mass memory so far is the federal bureaucracy, the "true possessors of institutional knowledge." Before Lesko's work, the government constituted a largely unexplored information source. This work fills the exploration gap well. As he points out, to call government workers "bureaucrats" is really to misname them. "Some 710,000 of these government workers are really information specialists. As taxpayers we pay their salaries and fund their research. As information consumers, we are entitled to a return on our investment." Let's put our government on-line, where we can get at it more easily, shall we? Until then, here's your Information Age map to Washington, D.C.

Ries, Al, and Jack Trout. *Positioning: The Battle for Your Mind.* New York: McGraw-Hill, 1981.

Ries and Trout spill all in this handbook for anyone who has a need to market something: self, idea, business or whatever. There's an excellent section on positioning yourself and your career. You can also clean out your own data banks by determining which companies have established niches in *your* brain. Weed them out ruthlessly: there's no need to continue to think in stereotypes when it comes to new products and services.

3.5.1
ON-LINE HISTORY

The number of on-line services is predicted to quadruple in the next five to ten years. This means an ancillary business of directories, magazines, reviews, tutorials, specialized software for accessing different kinds of services, and so on. It's going to be tough, for a while. Many on-line databases are very expensive to use and to learn how to use. Others are easier to use, but less useful in terms of the information they contain. There are no easy answers that a netweaver has to give to someone who asks, "Which on-line database should I use?" or even, "Which is the best on-line database directory?"

To start, you might want to check out the *Directory of Online Databases,* from:

Cuadra Associates, Inc.
2001 Wilshire Blvd., Suite 305
Santa Monica, CA 90403

A lot of the existing databases are available through Lockheed's Dialog service. You are charged according to which database you use, and the charge is pretty hefty, so you'd best know what you're after and how to get it before you dial up Dialog. However, they also have a lower-cost service, available in off-hours, called Knowledge Index, which is more home-consumer- and small-business-oriented, and a lot cheaper. Get information about either from:

DIALOG Information Services, Inc.
3460 Hillview Avenue
Palo Alto, CA 94304
415-858-3785
800-227-1927
Telex 334499 Dialog
TWX 910-339-9221

**3.5.2
BUSINESS
INTELLIGENCE
INCREASE**

Synnott, William R., and William H. Gruber. *Information Resource Management: Opportunities and Strategies for the 1980s.* New York: John Wiley & Sons, 1981.

This is a MIS (Management Information Systems) bible for those working inside the corporation. Users are a group to be "cared for and fed," perhaps like domesticated cattle. On the whole, though, this book can come in handy for planning a career in some aspect of telecommunications or netweaving. One of their strategies: Human Motivation Seminars within the company. Purpose: "guiding staff members to a better understanding of themselves and others to improve interpersonal work relations." *Career paths* are slotted into strict compartmentalized end states: data processing, systems development, systems research, telecommunications, and "other." A useful organizing tool for more conservative organizations.

Meadow, Charles T., and Pauline A. Cochrane. *Basics of Online Searching.* New York: John Wiley & Sons, 1981.

Let's say you have installed a micro and a modem, and you want to begin tapping the enormous existing store of on-line databases. After learning how to get connected to the various systems, you'll have to begin to learn how to get them to tell you what you say you need to know. This is where on-line search strategies come in. To use the various *query languages*—languages designed to help the user access data—without spending an arm and a leg on connect time will require that you spend some time psyching out the various systems. This overview will help you get the most out of on-line services, either for your own business or for a client. It is just possible that in years to come such preliminary work will be eliminated, and on-line systems will teach you what you need to know without costing you a lot of money.

Dyer, William G. *Strategies for Managing Change.* Reading, Mass.: Addison-Wesley, 1984.

Just as increased personal intelligence means greater awareness of oneself and one's processes, so, too, organizational intelligence includes raising consciousness about its processes. This is called insight. As the services sector of the economy—consultants, organizational development people, educators, ministers, social workers, and the like—merges with the information sector, planned change will become the rule rather than the exception. Dyer provides, in an easily read 200 pages, the essential strategies and models used by successful change agents, helpers and managers. You can increase the likelihood of having an impact by constantly increasing your store of strategies and honing them to the status of fine tools.

3.5.3
OFFICE
AUTOMATION

Marill, Thomas. "Why the Telephone Is on Its Way Out and Electronic Mail Is on Its Way In." *Datamation,* August 1979, pp. 185–188.

This representative article on "OA" presents the MIS (Management Information System) slant in approaching microcomputerization. The problem with most of these analyses is that they assume that, when introduced by the "wrong" segment of a corporation, MIS managers and data processing departments will somehow "lose control" of the corporation's "information assets." In actual practice, if the micros are introduced by MIS and DP departments electronic message systems tend to be ignored and not used by the organization's participants. Problems are legion. The experts say the majority of these systems are too hard to learn how to use, which is why they are not adopted or adopted slowly. They are too hard to use because they have been designed by DP-oriented people for other MIS managers. The cult continues.

3.6
STANDARDS AND
STANDARDS
ORGANIZATIONS

Government Agencies and Resources Concerned with Telematics. Many governmental bodies study and oversee telecommunications nationally and internationally. The following listings are good sources for finding out what's going on and keeping abreast of developing legislation. Each of the committees and subcommittees of Congress will usually have a chairperson and staff director. Holders of these positions change from year to year as elections dictate. Calling or writing will net you the current persons in charge.

Subcommittee on Telecommunications, Consumer Protection and Finance
c/o House Committee on Energy and Commerce
2125 Rayburn House Office Building
Washington, DC 20515
202-225-2927

This subcommittee has been a major influence on the direction of deregulation in the telecommunications industry. As chairperson of the subcommittee, Rep. Timothy E. Wirth conducted a comprehensive study on the state of competition in the industry. The subcommittee is also concerned with questions relating to foreign competition and public information sources such as public radio and television.

Subcommittee on Human Rights and International Organizations
c/o House Committee on Foreign Affairs
2170 Rayburn House Office Building
Washington, DC 20515
202-225-5021

Subcommittee on Government Information, Justice, and Agriculture
c/o House Committee on Government Operations
2157 Rayburn House Office Building
Washington, DC 20515
202-225-5051

The Subcommittee on Government Information, Justice, and Agriculture is concerned to some degree with issues of freedom of information and privacy in telecommunications, national and international.

Subcommittee on Space Science and Applications
and
Subcommittee on Science, Research and Technology
c/o House Committee on Science and Technology
2321 Rayburn House Office Building
Washington, DC 20515
202-225-6371

Office of Technology Assessment (OTA)
600 Pennsylvania Avenue, S.E.
Washington, DC 20510
202-226-2115

The OTA was created in 1972, just around the time that ecology was gaining vogue. The mandate of the OTA is to provide Congress with well-researched analyses of public policy issues that may arise out of new scientific and technical developments. It studies what Congress asks it to study. It is especially concerned with information and communications technologies in all their aspects.

Subcommittee on Communications
c/o Senate Committee on Commerce, Science and Transportation
5202 Dirksen Senate Office Building
Washington, DC 20510
202-224-5115

Senator Barry Goldwater, a ham radio licensee for many years, is chairman of the influential Subcommittee on Communications as of this writing. The subcommittee is broadly concerned with communications, space, and consumer affairs with regard to both.

National Aeronautics and Space Administration (NASA)
400 Maryland Avenue, S.W.
Washington, DC 20546
202-755-2320

Space and telecommunications are intricately intertwined. Space policy is a concern of anyone who is interested in telecommunications development.

National Technical Information Service (NTIS)
5285 Port Royal Road
Springfield, VA 22161
703-487-4600

NTIS is part of the Commerce Department. It is an excellent source of technical information paid for by the taxpayers through federal research. NTIS also culls foreign technical reports and republishes them in the United States. The agency has plans for the "formation of a worldwide network for the interchange of scientific and technical information."

National Telecommunications & Information Administration (NTIA)
14th Street and Constitution Avenue, N.W.
Washington, DC 20230
202-377-1840

NTIA is also part of the Commerce Department. It assists in the formulation of international communications policies. It oversees such things as the management of international radio frequencies and the application of satellite technologies. Transnational data flow, privacy, and technical standards also come under its purview.

**3.7
POLITICS AND
TELEPOLIS**

Watzlawick, Paul. *How Real Is Real? Communication, Disinformation, Confusion.*
New York: Random House, 1976.
There are grounds for great paranoia here: the book demonstrates how communication using spoken, written, sign, and body language creates what we call "reality." People can literally drive each other crazy, deliberately (in the case of "intelligence agencies" distributing disinformation) or unintentionally, as in disturbed family systems. This basic book on communications is indispensable, and good immunization.

Bezold, Clement. *Anticipatory Democracy: People in the Politics of the Future.*
New York: Random House, 1978.
Anticipatory Democracy (or A/D) is a form of practical futurism. Combining participation with long-range goal-setting and planning in diverse settings, it suggests the processes of telepolis through examples taken from real experiences in A/D.

Relyea, Harold C. *The Presidency and Information Policy.* Washington, D.C.:
Center for the Study of the Presidency, 1981.
Security: national and otherwise. Information: public and private. The president: executive privilege and personal papers. This volume of essays surveys the ground of the issues without really reaching any conclusions. We have a government by leak and press release, official secrets, and dossiers. Netweavers interested in national information policy will find this an essential source document for finding out what those in the centers of power think the issues are.

**3.7.1
BANKING AND
CREDIT**

Martin, James. *Telematic Society: A Challenge for Tomorrow.* Englewood Cliffs,
N.J.: Prentice-Hall, 1981.
Martin is a leading theoretician of the Information Age, having traversed the transition from hardware (IBM) to pure information: he has published twenty-eight books on telecommunications and computers. This book was formerly titled *The Wired Society,* and it covers much of the same ground as Nora and Minc [Appendix III: 3.8.4]. He spends a great deal of time in this work talking

about banking and money, but nowhere does he provide us with a good definition of "wealth," which he talks about a lot in connection with banking. He, like many telegurus of yesteryear, is shy when it comes to dealing with concepts like "wealth" and "money" in the Information Age. His technology is down, but his social thinking and social criticism, if they can be called that, stop short of the basic redefinitions that the technology seems to demand we make. In any case, I heartily recommend reading any of his books because he's always lively and entertaining, and his books always are packed with solid information from cover to cover.

**3.7.2
MACROSYSTEM
CONFLICTS**

You can get a guide to local exchange fixed costs on a state-by-state basis from:

> Telecommunications Users Coalition
> 1300 N. 17th Street
> Arlington, VA 22209

This can help you in dealing with local Public Utility Commissions—now charged with regulating the "minimonopolies" created by the regional split in the phone company. It can also help you predict which regions are going to have the highest increases in local rates over the next few years. The TUC is primarily a business-oriented end-user lobby and information group.

Self, Robert. *Long Distance for Less.* New York: The Telcom Library, 1983.
    This is a guide to alternative phone companies for business users. Available from:

> The Telcom Library
> 205 W. 19th Street
> New York, NY 10011
> 212-691-8215

Publications about the teletremors in the economy, and a computer program listing rates and tariffs are available from:

> Economics and Technology, Inc.
> 101 Tremont Street
> Boston, MA 02108

**3.8
GLOBAL ACCESS**

The Planetary Initiative for the World We Choose, or just Planetary Initiative, is an international effort to apply the principles of anticipatory democracy to global issues. Local community forums have been started all over the country. For more information, contact:

> Planetary Initiative
> 3025 Arizona Street
> Oakland, CA 94602

Issues in international information, transnational data flow and communications are followed closely by The Media Institute in Washington, DC. It is one of the primary advocates of "the free flow of information across international borders." As such, the institute represents the interests of existing media enterprises: broadcasters, advertisers, news agencies, and so on. Global micro communications—in which *individuals* participate in the "free flow"—is definitely not on the institute's agenda. Nonetheless, netweavers who want to participate in global processes should monitor the output of this organization.

The Media Institute
3017 M Street, N.W.
Washington, DC 20007

**3.8.2
SATELLITES,
SPACE,
AND DBS**

Easton, Anthony T. *The Home Satellite TV Book: How to Put the World in Your Back Yard.* San Francisco: Wideview Books, 1982.

And you thought you had trouble keeping the neighbors' kids out of the pool! It's all here, or at least most of it: DBS ("birds"), Mini-CATV systems, setting up satellite TV dealerships, and more. An excellent first source document for video and satellites.

**3.8.3
VIDEO REVOLT**

Maltin, Leonard. *The Whole Film Sourcebook.* New York: New American Library, 1983.

In case you hadn't noticed, film and video are merging, and both are going digital. Not a moment too soon, either, because we're losing a lot of our early filmic heritage to entropy as priceless historic prints fade and crumble. Maltin has done an excellent job defining the film information territory. Any netweaver interested in raiding for goodies will find this source tantalizing: film courses, TV, cable, video, books on film, grants, exhibitors, distributors, and on and on.

Hurst, Walter E., Johnny Minus, and William Storm Hale. *Your Introduction to Film, TV, Copyright, Contracts and Other Law.* Hollywood: Seven Arts Press, 1976.

The information environment generates lawsuits almost as prolifically as it generates new media products and services. Media attorneys specializing in film and television law say that their biggest frustration with clients is that these "creative types" refuse to learn enough about the laws to keep out of trouble, nor enough to understand the legal advice they may get once they get into trouble. If you plan to synthesize your micro and networking experience with any aspect of the visual media, this book will help you avoid trouble. Seven Arts Press is at 6605 Hollywood Blvd., Hollywood, CA 90028.

For a newsletter aimed at film/video/radio/TV producers and the general public, write:

National Alliance of Media Arts Centers
5 Beekman Street, Room 600
New York, NY 10038

**3.8.4
THE FRENCH**

Nora, Simon, and Alain Minc. *L'Informatisation de la société.* Paris: La Documentation Française, 1978. *The Computerization of Society.* Cambridge, Mass.: M.I.T. Press, 1981.

This basic resource document delineates the macrosystems questions and conflicts that have to be addressed by national governments in the Information Age. It also suggests strategies for individuals and organizations up to the corporate level. We have to believe that, at some level of its pyramid, IBM has absorbed this document in its international marketing strategy. This book was originally written as a report to the president of France. It recommends a *national plan,* which, curiously, doesn't exist yet, here or in France.

**3.9
NETWEAVING**

Hiltz, Starr Roxanne, and Murray Turoff. *The Network Nation: Human Communications via Computer.* Reading, Mass.: Addison-Wesley, 1978.

A classic in the field of communications and computing. A mainframe of reference is employed throughout. Hiltz and Turoff based much of this work on their experience with EIES (Electronic Information Exchange Service) and in government work. EIES still exists. Imagine an enormous university campus cocktail party, with the pseudo-erudite and hangers-on rubbing shoulders with Nobel Prize winners. Now imagine them gathering in corners in groups to gossip about their special interests ("conferences"). Now imagine yourself as an eavesdropper at this party. Along with birthday congratulations and book discussions and interpersonal chitchat, you hear snatches of conversation about just about everything under the sun. Now imagine that you had to *pay* to attend this cocktail party and contribute your own infinitely more profound comments to the ongoing conversation. That's EIES. Even though some feel there's better conversation to be had for less money in their own community networks, EIES and its history have influenced a great number of early netweaving pioneers, including me. This book belongs on your shelf of computing history.

**3.9.1
PERSONAL
LEARNING
NETWORKS**

Just the process of learning how to access a local network will begin to teach you about telecommunications. From there, much depends on how self-motivated you are and how much formal structure you require to learn well. I am aware of at least two telecommunications-based learning programs so far. One is through a traditional university combined with a mass-based information utility (the Source). The program is limited. The other is through a specially set-up network using proprietary software and protocols. You have to buy the software package as part of the "tuition" to the school. On the whole, you are better off scouting your community for the expertise you need (the "teachers") and setting up your own learning network on any particular subject you happen to be interested in. Of course, if your local community doesn't have the expertise available, for whatever reason, then the next step is to investigate commercially offered telecourses. Don't forget that public television stations have been offering "educational television" for college credit for some years now. You might be able to set up a co-media program, whereby you have access to the knowledge base through both television and a micro network. There is still lots of room for innovation in this area, which is why I'm not ready to list any of the existing "network-based schools" here as yet. Maybe next year.

However, there is much to be learned from the games networks. And well-constructed games can teach an enormous amount. I recommend that colleges and universities explore the possibilities inherent in "games for credit." To begin with, you can learn about what's possible *today* by getting on board an existing games network and exploring. Here's a place to start:

GameMaster
1723 Howard Street, Suite 219
Evanston, IL 60202

GameMaster is available for micro access. Its network is managed by and its games programs are run on an AlphaMicro computer that allows many callers at the same time. All kinds of on-line games have been started: strategic, sports, single-player arcade-type games, and more. If you have an interest in the possibilities as either a designer or user, this is a premier network.

**3.9.2
REMOTE
SUPERVISION/
COOPERATION
NETWORKS**

Lindenfeld, Frank, and Joyce Rothschild-Whitt, eds. *Workplace Democracy and Social Change.* Boston: Porter Sargent Publishers, 1982.

A quality circle in a factory [Appendix III: 3.4] implements more democracy in the workplace. Granted that "democracy" is a moving target, managers of traditional capitalist organizations are probably not really interested in the ideology involved: the fact is that decisions made by those who must implement them are usually not only better decisions—less costly, time-consuming, and so on—but more likely to be implemented. This is a pragmatic reason for incorporating more workplace democracy all the way around, at all levels of existing organizations. If you find yourself building a business, this book is full of practical advice, although some of it is framed in obsolete terms.

**3.9.4
COMMUNITY
SOAPBOX
NETWORKS**

Lassey, William R., and Richard R. Fernandez, eds. *Leadership and Social Change.* La Jolla, Calif.: University Associates, 1976.

If you are attempting to introduce new technology and new methods to a small town or suburban community, you are, wittingly or unwittingly, placing yourself in a position of leadership. You are attempting to introduce social change. This book of readings may help you make to make explicit and conscious the leadership skills you already have, or think you have. It is more theoretical than most, insofar as it covers the various "theories of leadership" and the history of the concept of leadership, but is useful nonetheless. Along with *Synergic Power* [Appendix III: 3.9.7], this book can help your grope your way through many kinds of institutions in government, business, and towns.

**3.9.5
CONSULTANTS
AND
CONSULTANTS'
NETWORKS**

Kerzner, Harold. *Project Management for Executives.* New York: Van Nostrand Reinhold, 1982.

Why do plans fail? Why do some networks materialize and not others? Why do some start out with promise, then falter and die? A lot probably has to do with the influence of the original organizers and their planning expertise or lack thereof. If you keep this reference at hand during your project planning, you're bound to have a head start on succeeding in getting your new network(s) off the ground. If you work for yourself, or a small organization, or a larger corporation, and you handle information, think of yourself as an executive for purposes of managing your information territory.

3.9.6
SPECIAL-INTEREST
NETWORKS (SPIN)

Those who provide electronic information services constitute a SPIN: they are fond of calling themselves the "information providers." All hail those who provide! How does this special interest communicate? With a *print-and-paper-based newspaper,* naturally. *Information Today* covers the same territory as a lot of other newsletters, magazines, and on-line services, but it costs a lot less than many of them. From:

> Information Today
> Learned Information, Inc.
> 143 Old Marlton Pike
> Medford, NJ 08055

Apple Computer has a grant program going called Community Affairs or Community Networking. This program *gives* micros to local community groups who wish to start an information exchanging and cooperation network. Two or more nonprofit community organizations apply for the Apple grants. Those accepted get complete micro systems, including modems and software for on-line use. For more information on how to apply, restrictions and conditions involved, contact the Apple Community Affairs Program, in care of Apple. (Address in Appendix IV: 4.12)

3.9.7
DISTRIBUTED
NETWORK
ORGANIZATION

Craig, James H., and Marge Craig. *Synergic Power: Beyond Domination and Permissiveness.* Berkeley, Calif.: Proactive Press, 1974.
    In any given relationship, but especially in organizational ones, you may find yourself in one of the corners of the power triangle:

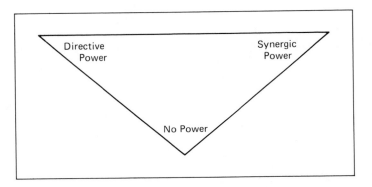

If you are a typical person, you probably think of yourself as being more in the "no power" corner of the triangle. The Craigs show you how to use your desire to work with others effectively to forge a set of political tools. Moving into the synergic power corner means giving up nothing except the desire to dominate, manipulate or coerce others into doing your will. This is by far one of the best systems-oriented organizing manuals I have ever come across. And it is ideal for those wishing to create networks, because its principles can be articulated very well in the network-based relationships that must emerge if the network is to be successful at getting its "job" done, whatever the job might be.

**3.9.8
PERSONAL
GLOBAL
NETWORKS**

Katzan, Harry S., Jr. *Multinational Computer Systems: An Introduction to Transnational Data Flow and Data Regulation.* New York: Van Nostrand Reinhold, 1980.

A good overview of the technical, political, and economic issues involved in global networking as it has been so far. Provides the necessary background before you jump in and start trying to "organize a global network" at whatever level.

Once you *do* get started with your global organizing, you can keep in touch with others doing the same thing through:

Mutual Inquiry Network Development (M.I.N.D.)
Box 14431
San Francisco, CA 94114

APPENDIX IV

Building
and Weaving

4.1 MILESTONES

Axiom: Everyone can't know everything about everything.
Corollary: But some people can fake it pretty well.

An Age of Innovation: The World of Electronics 1930–2000. By the Editors of *Electronics* magazine. New York: McGraw-Hill, 1981.

Antebi, Elizabeth. *The Electronic Epoch.* New York: Van Nostrand Reinhold, 1982.
    Both these lavish volumes provide a good recent history of developments in electronics and modern physics, including good sections on the origins of digital computers. Both are profusely illustrated and include technical explanations of the various milestones in the evolution of circuitry on planet Earth—for the layperson. Highly recommended.

Lavington, Simon. *Early British Computers.* Bedford, Mass.: Digital Press, 1980.
    Prior to 1940, a "computer" was a clerk, using a hand-operated calculating machine, who "computed" such things as actuarial tables, ballistic trajectories, and the like. The evolution of the computer owes as much to the British as to Americans. This overview details the work done in the United Kingdom between 1935 and 1955. The British are credited with pioneering "the world's first stored-program computer, the first commercially available computer, and the first transistorized computer." So there.

Jastow, Robert. *The Enchanted Loom: Mind in the Universe.* New York: Simon & Schuster, 1981.
    From Jastrow: "If scientists can decipher a few of the brain's signals today, they should be able to decipher more signals tomorrow. Eventually, they will be able to read a person's mind.
    When the brain sciences reach this point, a bold scientist will be able to tap the contents of his mind and transfer them into the metallic lattices of a computer. Because mind is the essence of being, it can be said that this scientist has entered the computer, and he now dwells in it."

The rest of *The Enchanted Loom* is equally provocative, entertaining, and informative.

**4.2
ENTRY LEVEL:
PERSONAL
SOFTSPACES**

Sigel, Efrem, ed. *Videotext: The Coming Revolution in Home/Office Information Retrieval.* New York: Harmony Books, Crown Publishers, 1980.

Tydeman, J., H. Lipinski, R. Adler, M. Nyhan, and L. Zwimpfer. *Teletext and Videotext in the United States: Market Potential, Technology, Public Policy Issues.* New York: McGraw-Hill, 1982.

When it comes right down to it, the use of the terms "teletext" and "videotext" are merely ways of recategorizing old and new technologies in ways that reinforce industrial (Second Wave) associations. Thus microcomputer-based systems are lumped in with special "black box" schemes for delivering intangibles that the public will—it is hoped—pay for. Both of these books are good examples of this approach to the telecommunications transformation. In addition, the July 1983 issue of *Byte* magazine (vol. 8, no. 7) bows in this direction with a themed "Videotext" issue.

The main question should be, for any potential user, "Why do I need this?" If you can come up with positive reasons, then take the next step and look at microcomputer-based answers to the same question.

AT&T (American Bell) doesn't seem to be worried about the crash of the phone system due to increased data communications. American Bell sees itself as best able to deliver videotext services. According to one of American Bell's videotext marketing managers, "the telephone lines are presently the best suited to deliver videotext because: (1) Cable TV is in fewer homes across the country than telephone lines, and (2) there is presently little videotext activity going on in the cable industry because the hybrid products that the Cable Industry needs aren't presently available" (correspondence with the author, November 1983).

Thus, American Bell is proceeding full speed ahead in the implementation of its Sceptre videotext system. Rather than seeing both the production and the consumption of information as two sides of the same softspace, Behomoth Moguls are widening the split between those who can access and create new information and those who can merely access it. The Sceptre "frame generation system," essentially a microcomputer system for creating graphic-and-text "frames" of information, is being sold to the "providers" for $40,000 to $70,000. It is slow and cumbersome to use. The graphics aren't so good, at that, compared with graphics created with even inexpensive home computers. Their access terminal, on the other hand, is slightly more sophisticated than a pocket calculator with a built-in 1,200-baud modem and a "black box" for marrying the consumer's home television set with the consumer's telephone. The consumer doesn't get to do much more with this hardware than access what Ted Nelson called "the greatest schlock for the greatest number" ("Text Media for Tomorrow," *High Technology,* May 1983, p. 64). Meanwhile, microcomputer manufacturers and the makers of micro peripheral equipment are gearing up to equip micros with the ability to decode videotext frames, including those that will be generated by American Bell videotext systems.

The Videotext 1983 convention—where those identifying themselves under the videotext banner display themselves—had about fifty-five vendors of hardware and software and information, and owners of other conduits such as cable.

IBM is trying to sell its "videotext conferencing" developed for IBM customers who already have their Series 1 computers, and those who can be persuaded to buy one just for videotext. IBM is hedging its videotext bets by supporting both the PRESTEL (European) and NAPLP (North American, also supported by American Bell) standards. The IBM-PC outfitted with a Wolfdata card creates the videotext terminal with which to access these services.

Keycom's videotext service was due to begin in early 1984. They want to provide cheapie terminals based on pop-computers, including TRS-80s, Apples, and others.

Sony and Panasonic provide NAPLP [4.9.5] videotext decoders with some of their lines of television sets. Nearly all the black box decoders offered or being planned have an RS442 port for attaching printers, so users will be expected to buy printers along the way.

Nearly all of them support the NBS Data Encryption Standard [4.10.4], which is a downer for those interested in protecting civil liberties from big government and big business.

**4.2.3
DATA
COMMUNICA-
TIONS**

If you plan to work with networks, modems, and people as part of your career, a subscription to *Data Communications* will prove worthwhile. Not for the casually interested, and certainly not necessary just to get your own softspace in order and connected to the networks, *DC* is to netweavers what *Byte* is to computer enthusiasts.

> *Data Communications Magazine*
> McGraw-Hill Publications Co.
> 1221 Avenue of the Americas
> 42nd Floor
> New York, NY 10020

**4.3
THE TELEPHONE
SYSTEM(S)**

Kuecken, John A. *Talking Computers and Telecommunications*. New York: Van Nostrand Reinhold, 1983.

In addition to presenting a concise history of early telephony, this reference demystifies the techniques behind changing voices to digital information and back again. The latter includes one of the best presentations of acoustics and speech digitization I've ever seen. Highly recommended.

**4.3.5
CHOOSING AN
ALTERNATIVE
PHONE SERVICE**

You can keep track of who's who in the alternative phone game by contacting:

> Association of Long Distance Telephone Companies
> 2000 L Street, N.W. #200
> Washington, DC 20036
> 202-463-0440

Each of the following companies offers a mix of services to home and business users. Rates and pricing structures change frequently, so it is recommended that you contact the representatives in your city. Get local numbers from:

SPRINT
Southern Pacific
  Communications (SPC)
One Adrian Court
Burlingame, CA 94010
800-521-4949
800-645-6020 (Mich.)

MCI
MCI Building
17th & M Streets, N.W.
Washington, DC 20036
800-521-8620
800-482-1740 (Mich.)

U.S. Transmission System's
  "Longer Distance" (ITT)
U.S. Transmission Systems, Inc.
P.O. Box 732, Bowling Green
  Station
New York, NY 10004
800-438-9428
212-797-2511 (N.Y.)

Western Union's "MetroFone"
Western Union
1 Lake Street
Upper Saddle River, NJ 07458
800-325-6000

Satellite Business System's
  "Skyline"
John Marshall Building
8283 Greensboro Drive
McLean, VA 22102
800-698-6900

Allnet
Combined Network, Inc.
100 South Wacker Drive
7th Floor
Chicago, IL 60606

We have had good experiences with both Sprint and MCI, although some users we know complain about both. The suspicion is that quality of service may vary depending on where you're located.

Simon, Samuel A., and Joseph W. Waz, Jr. *Reverse the Charges: How to Save Money on Your Phone Bill.* New York: Pantheon Books, 1983.
  Among a host of new books on the telephone company, this slim one serves well. Though not, I hope, the last organization of its kind, the Telecommunications Research and Action Center (TRAC) is a consumer-oriented nonprofit agency, and the source of this reference for the telebewildered. They also publish *Access,* a monthly journal on media-related subjects.

TRAC
Box 12038
Washington, DC 20005

Say you want to set up a network consisting of several nodes in different cities. Phone calls for data transfers will have to take place on a daily basis. How do you estimate the costs for phone connections alone, assuming you want a private line to link the nodes of your network? For that kind of planning and budgeting, there is a master catalog of all the possible communications channels you might use, their costs and limitations. It's called the *IRH Intercity Rates and Services Handbook.* It comes in two volumes: (1) Private Line Services: analog channels, digital, wideband, satellite, and low-speed data channels. (2) Measured

Use Services: voice network, voice point-to-point, data service, facsimile, tele-type, telegraph, and electronic mail.

This is an information tool for telecommunications professionals, so it's costly. Information from:

Economics and Technology, Inc.
101 Tremont Street
Boston, MA 02108
617-423-3780

4.3.7
TOUCH TONE OR
PULSE DIAL
EQUIPMENT

You can get a phone book of phones from:

Comdial
Consumer Communications Division
9620 Flair Drive
El Monte, CA 91731
800-854-0561
800-432-7257 (Calif.)

Comdial carries the Kermit the Frog Phone, the Darth Vader Speakerphone, and others made by the American Telecommunications Corporation. Comdial also has phone accessories, such as modular jack wall plates, extra cords, double modular ("Y") adaptors, and more.

4.5
SIGNALS

Nichols, Elizabeth A., Joseph C. Nichols, and Keith R. Musson. *Data Communications for Microcomputers.* New York: McGraw-Hill, 1982.

If you find yourself thirsting after more technical data than is contained in this sourcebook, try the above paperback to continue your quest. It is one of the most technically useful of the available technical books on data communications. Unfortunately, many of the books containing "data communications" or "telecommunications" in their titles do little more than present warmed-over printouts from mass-based data banks. This book does not err at all in this direction.

4.7
SERIAL VERSUS
PARALLEL
TRANSMISSION

A connection between the serial ports of two micros is probably the easiest sort of linkage you can make. However, serial is as serial does. Just because the machine's manual specifies a particular outlet as a "serial port" does not mean that it will connect easily to another machine similarly labeled. Would that it were so simple!

To be absolutely sure, the two systems should have RS-232C serial ports. Connecting any other two "serial ports" could blow out either or both machines. Sizzled brains is not the object here! For more on this, see section [4.16].

4.7.1
SERIAL CARDS,
UARTs, USARTs,
AND SIOs

In addition to UARTs, there are other integrated circuits now being used in modems and serial cards: USARTs (Universal Synchronous/Asynchronous Receiver-Transmitter) and SIOs (Serial Input/Output circuits). All these circuits use *shift registers* to convert parallel bytes to serial ones and back again.

**4.7.2
CHARACTER-
ORIENTED
PROTOCOLS**

Translation Tables.   There are keyboards and there are keyboards. If you have an IBM PC, you'll have function keys and a numeric keypad and a variety of other keys that a TRS-80, by contrast, will not have. Likewise, while ASCII is a standard, and while most keyboards on micros and low-cost terminals claim to be ASCII, that may only be 99 percent true in a given case. Some keyboards will simply not be able to generate some of the ASCII characters you will need in telecommunications work.

In addition, there may come a time when you want to communicate with a baudot terminal, or some other kind of exotic creature on the nets.

This becomes more important when you are linking up with your larger office computer from home, for instance. You may want to generate certain ASCII characters and transmit them to the big machine in order to execute some of the mainframe's functions. That's the only way you'll get it to do what you want.

Even though your micro by itself may not be capable of generating all the ASCII characters you need, you will probably be able to pick up software that can handle the job for you.

All you have to do is make sure that your communciations software provides *translation tables* or the ability to generate such tables. What the translation table does is take characters or character sequences that your keyboard *can* generate and turn them into strings of bits standing for characters that are missing. It will then send these down the wire to the other computer. The receiving end, needless to say, couldn't care less whether or not your machine can display such characters, much less generate them, so when the duplex echo comes back, you may get a weird character on your screen. No matter. It's the job we want done that's important here.

Translation tables can work the other way around, of course, and take incoming characters from the remote computer and translate them into characters that your micro can use or display easily. When and if you get around to linking up with typesetting machines, for example, translation tables will help you convert your micro's ASCII characters into control sequences that the typesetting machine can use in its machinations on the way to paper.

Another use for such tables is between two dissimilar word processor text files. Say you want to send a file created by an Ajax word processor to a friend who runs an Acme W/P. A translation table in your communications program can take care of changing or getting rid of any special control characters that can only be used with your word processing program. On the other end, your friend may have an incoming table to translate text into files suitable for her micro's text handling system.

**4.7.2.1
ASCII**

There are many different ways of presenting the ASCII characters in chart form. Some have the advantage of saving space on the page but may be difficult to read for those not used to solving technical puzzles. The following list is simply the ASCII character set with accompanying codes. Extended ASCII character sets incorporate the 128 standard symbols shown below. However, they also use eight bits instead of seven, and the meanings of the extra 128 characters that result will vary from machine to machine and from software to software.

Character	Binary Bits 6–0	Hex	Decimal
NUL	0000000	0	0
SOH	0000001	1	1
STX	0000010	2	2
ETX	0000011	3	3
EOT	0000100	4	4
ENQ	0000101	5	5
ACK	0000110	6	6
BEL	0000111	7	7
BS	0001000	8	8
HT	0001001	9	9
LF	0001010	A	10
VT	0001011	B	11
FF	0001100	C	12
CR	0001101	D	13
SO	0001110	E	14
SI	0001111	F	15
DLE	0010000	10	16
DC1	0010001	11	17
DC2	0010010	12	18
DC3	0010011	13	19
DC4	0010100	14	20
NAK	0010101	15	21
SYN	0010110	16	22
ETB	0010111	17	23
CAN	0011000	18	24
EM	0011001	19	25
SUB	0011010	1A	26
ESC	0011011	1B	27
FS	0011100	1C	28
GS	0011101	1D	29
RS	0011110	1E	30
US	0011111	1F	31
SP	0100000	20	32
!	0100001	21	33
''	0100010	22	34
#	0100011	23	35
$	0100100	24	36
%	0100101	25	37
&	0100110	26	38
'	0100111	27	39
(	0101000	28	40
)	0101001	29	41
*	0101010	2A	42
+	0101011	2B	43
'	0101100	2C	44
- (hyphen)	0101101	2D	45

Character	Binary Bits 6–0	Hex	Decimal
.	0101110	2E	46
/	0101111	2F	47
0	0110000	30	48
1	0110001	31	49
2	0110010	32	50
3	0110011	33	51
4	0110100	34	52
5	0110101	35	53
6	0110110	36	54
7	0110111	37	55
8	0111000	38	56
9	0111001	39	57
:	0111010	3A	58
;	0111011	3B	59
<	0111100	3C	60
=	0111101	3D	61
>	0111110	3E	62
?	0111111	3F	63
@	1000000	40	64
A	1000001	41	65
B	1000010	42	66
C	1000011	43	67
D	1000100	44	68
E	1000101	45	69
F	1000110	46	70
G	1000111	47	71
H	1001000	48	72
I	1001001	49	73
J	1001010	4A	74
K	1001011	4B	75
L	1001100	4C	76
M	1001101	4D	77
N	1001110	4E	78
O	1001111	4F	79
P	1010000	50	80
Q	1010001	51	81
R	1010010	52	82
S	1010011	53	83
T	1010100	54	84
U	1010101	55	85
V	1010110	56	86
W	1010111	57	87
X	1011000	58	88
Y	1011001	59	89
Z	1011010	5A	90
[	1011011	5B	91
\	1011100	5C	92
]	1011101	5D	93

Character	Binary Bits 6–0	Hex	Decimal
	1011110	5E	94
—	1011111	5F	95
	1100000	60	96
a	1100001	61	97
b	1100010	62	98
c	1100011	63	99
d	1100100	64	100
e	1100101	65	101
f	1100110	66	102
g	1100111	67	103
h	1101000	68	104
i	1101001	69	105
j	1101010	6A	106
k	1101011	6B	107
l	1101100	6C	108
m	1101101	6D	109
n	1101110	6E	110
o	1101111	6F	111
p	1110000	70	112
q	1110001	71	113
r	1110010	72	114
s	1110011	73	115
t	1110100	74	116
u	1110101	75	117
v	1110110	76	118
w	1110111	77	119
x	1111000	78	120
y	1111001	79	121
z	1111010	7A	122
{	1111011	7B	123
|	1111100	7C	124
}	1111101	7D	125
˜ (tilde)	1111110	7E	126
DEL	1111111	7F	127

Generating the familiar symbols on most micro keyboards is as easy as hitting the keys showing the symbol you want. But what about the special control characters? Most of those can be generated by using the CONTROL or CTR key—found on most keyboards—in conjunction with other keys. The spacebar, of course, is always a space (SP—ASCII 32). The control character abbreviations and their control key sequences are as follows:

Hold down the Control key and press:

[Shift] [P]	NUL—null, or all zeros
[A]	SOH—start of header or heading
[B]	STX—start of text
[C]	ETX—end of text

[D]	EOT—end of transmission
[E]	ENQ—enquiry
[F]	ACK—acknowledge
[G]	BEL—bell
[H]	BS—backspace
[I]	HT—horizontal tabulation
[J]	LF—line feed
[K]	VT—vertical tabulation
[L]	FF—form feed
[M]	CR—carriage return (or Enter key)
[N]	SO—shift out
[O]	SI—shift in
[P]	DLE—data link escape
[Q]	DC1—device control 1 (also the X-on character)
[R]	DC2—device control 2
[S]	DC3—device control 3 (also the X-off character)
[T]	DC4—device control 4
[U]	NAK—negative acknowledge
[V]	SYN—synchronous idle
[W]	ETB—end of transmitted block
[X]	CAN—cancel
[Y]	EM—end of medium
[Z]	SUB—substitute
[Shift] [K]	ESC—escape
[Shift] [L]	FS—file separator
[Shift] [M]	GS—group separator
[Shift] [N]	RS—record separator
[Shift] [O]	US—unit separator
[Del] or [Rubout]	DEL—delete

4.8 MODEMS   The microcomputer telecommunications world has two premier modem manufacturers: Hayes Microcomputer Products company in Atlanta, Georgia, and Novation, Inc., in Tarzana, California.

Hayes Microcomputer Products, Inc.
5923 Peachtree Industrial Blvd.
Norcross, GA 30092
404-449-8791

Novation, Inc.
18664 Oxnard Street
Tarzana, CA 91356
213-996-5060

If you use a Commodore VIC-20 or 64, you'll need a Commodore modem, as well. Commodore's address is in Appendix IV: 4.12.

Racal-Vadic, Inc. has been making modems for mainframe computers and large-scale networks for many years. This company also now carries high-quality modems for micro telecommunications use.

Racal-Vadic, Inc.
222 Caspian Drive
Sunnyvale, CA 94086

A low-cost, no-frills modem that works with most micros is available from Anchor Automation. Called the Volksmodem, this 300-baud workhorse has been praised by a wide variety of users. Information from:

Anchor Automation, Inc.
6913 Valjean Avenue
Van Nuys, CA 91406

**4.9
PROTOCOLS AND
STANDARDS**

Lest I get howls of protest from backers of Unix, I refer the interested netweaver to the August 1983 issue of *Systems & Software,* p. 11. In a well-put letter to the editor, a knowledgeable reader demolishes the idea that any of the current operating systems (OSs) will become the long-run standard. Parallel arguments can be made against adopting various other standards in the micro world, particularly in the area of floppy disk sizes and formats, of which there are now between ninety and one hundred. Look at it this way: is your nervous system composed of only one kind of neuron?

Quick and Dirty Settings. For the majority of your on line work, your protocol settings will be one of two *default* settings. These are shown in tabular form here. If you are not sure what protocols are being used by the system you're dialing up, try one or both of these. If these don't do the job, then you'll have to contact the parties who are running the system you're trying to connect up with and have them tell you what protocols you must use.

	Setting A	Setting B
Baud rate:	300 or 1,200	300 or 1,200
Data bits:	8	7
Duplex:	Full	Half
Parity:	None	Even
Stop bits:	1	None

Sometimes a system will ask you if you need *nulls* or *linefeeds* when you first log on. A null is a time-wasting, do-nothing ASCII character designed to mark time while your local system does what it needs to do to keep up with the transmitting system. Nulls are frequently used at the ends of lines to give your printer head time to return to the beginning of the next line and get ready to print a new line of text.

A linefeed is a character that tells your printer or terminal screen to go to the next line before printing out any more text. Otherwise the lines get printed over one another. Whether you need nulls or linefeeds will depend on your software, printer, and hardware capabilities. But it's easy to tell, in any case. If your printer drops characters at the beginning or ends of lines, ask for nulls. If it overprints lines, ask for linefeeds.

Many Remote CP/M systems (RCP/M: a bulletin board program for CP/M micros) use the MODEM7 protocol, a refinement of the XMODEM (Christensen) protocol. MODEM7 is also a terminal program using that same protocol, developed by Mark M. Zeiger and James K. Mills.

There are other protocols out and about. Three others that you might hear about or have to use now and again are REACH, COMMX, and AMCALL. The first is used on Health/Zenith systems, and the latter two on some CP/M systems. For more information, contact:

REACH
The Software Toolworks
15233 Ventura Blvd., Suite 1118
Sherman Oaks, CA 91403
818-986-4885

COMMX
Hawkeye Grafix
23914 Mobile
Canoga Park, CA 91307

AMCALL
Micro-Call Services
9655-M Homestead Court
Laurel, MD 20810

**4.9.1
HANDSHAKING**

The ASCII characters for X-on (DC1—ASCII $11) and X-off (DC3—ASCII $13) are handshaking conventions used to tell a transmitting system when to start and stop sending information. Many computer systems used this convention. Your software should support it.

**4.9.3
ISO NETWORK
REFERENCE
MODEL**

The International Standards Organization is a worldwide body, attempting to set worldwide standards. It is composed of representatives from standards bodies within various nations. The representative for the United States comes from ANSI, the American National Standards Institute.

You can get a list of various ANSI standards specifications by writing to:

American National Standards Institute
1430 Broadway
New York, NY 10018

For example, the ANSI specification labeled *X3.64* has been described as "the most comprehensive standard for information interchange yet devised." Net-weavers are encouraged to support this standard since in the long run it will ensure compatibility between diverse peripheral devices and prevent users from having to be locked to a single vendor whose protocols might be proprietary (unique to that company's devices).

**4.9.4
NETWORKING
STANDARDS**

The XMODEM program is one of the earliest communications terminal programs for micros. It uses the so-called XMODEM protocol, also called the Christensen protocol after the developer of both the XMODEM program and the protocol it uses. Ward Christensen also wrote—with Randy Suess, another hobbyist/pioneer—one of the first micro versions of so-called Bulletin Board Systems (BBSs). To this day, the XMODEM program is still available through publicly accessible micro networks. Both the program and its protocols run primarily on CP/M compatible micros and are worth support by netweavers. The protocols include error checking.

**4.9.5
NAPLP**

Sometimes this protocol is referred to as NAPLPS, the final *S* standing for Syntax. The exciting thing about NAPLPS graphics is not that the behemoth world begat it, but that it has applications beyond telebanking and teleshopping and other teledrivel. NAPLPS graphics combined with micro conferencing is an exciting way to communicate. This opens the way to teaching, graphic design, other kinds of industrial design, and all kinds of on line interactive systems that use color and shape in addition to letters and numbers. NAPLPS has been called "the most powerful graphics code ever developed." Is it worth supporting? Yes. It meets most of the major criteria for standards that netweavers concern themselves with:

- It is not a proprietary standard, yet it has been accepted by enough of the major communications companies around the world to maintain its viability and longevity.
- NAPLPS graphics are "resolution-independent." This means that, for example, if you have a display screen capable of high resolution, NAPLPS information will appear on your screen to take advantage of this capability. By the same token, low-resolution display devices will display all of the information *it* is capable of.
- The NAPLPS code requires only 10 percent of the memory space used by other graphics systems to display the same graphic information. NAPLPS is a "do more with less" encoding scheme.
- NAPLPS integrates graphics and text. It is designed to be integrated into other communications/display systems such as videotape, telephone, TV, and computer database.

For more information on NAPLPS and related micro graphics issues, see the February 1984 issue of *Computer Graphics World*.

Computer Graphics World
1714 Stockton Street
San Francisco, CA 94133
415-398-7151

**4.11.4
ECHOPLEX**

Note that both the local and remote micros must be set correctly—usually through software commands—to accommodate echoplex or no echoplex. When no echoplex—or half duplex—is in effect, then the transmitting micro will display the transmitted character on the screen as well as send it through the modem. If *both* systems are set for echoplex, then characters will appear twice on your screen. Your local system sends the character to your screen and to the remote, and the remote echoes it. The cure: set your system to half duplex. If *neither* system is set up for echoplex, then nothing will appear on your screen at all. The cure: set your machine on full duplex.

	Half Duplex	Full Duplex (echoplex)
Originate	Outgoing characters displayed locally. Incoming characters *not* echoed for remote display.	Characters neither displayed nor echoed.
Answer	Outgoing characters displayed locally. Incoming characters *not* echoed for remote display.	Displays locally *and* echoes incoming characters for remote display.

**4.12
MICROS,
TERMINALS, AND
MICROS AS
TERMINALS**

There is no perfect machine. All designs are compromises. Assume that communications is high on your list of priorities. From there, choosing a micro is a matter of putting together your personal package of options, then shopping for a machine to fit *your* specifications.

Is there an "objective" way to compare information systems? Well, yes and no. The best way to comparison-shop features is by means of criteria that you set yourself, after a thorough analysis of the larger system in which you intend to use the micro. This means that the criteria themselves will be "subjective," while your analysis can be "objective."

Professional consultants and designers have devised rules of thumb [1.2.2] for making such comparisons. You may find such an approach useful. For example, one method of comparing microcomputer systems is on a cost-per-function basis.

First, list the functions that are most important to you (if they're all important, include them all): word processing, telecommunications, graphics, and so on. These specifications can be expanded and refined, of course, as you do more research. Word processing could be detailed according to specific text-handling *features* you want on your system. In fact, the more refined you make your comparison, the more likely you'll arrive at a satisfactory decision.

Next, use your list of functional criteria to determine which hardware and software combinations will make the ideal real. This gets interesting quickly, since options proliferate and combinations can seem infinite. It may be that your list of features was too ambitious, coming in at a cost above what you are willing to consider. In any case, you'll arrive soon enough at a list of likely candidates. To compare them with each other, simply divide the total cost of the system by the number of provided features that are important to you. Each system you're looking at will then have a CPF (cost-per-function) that helps you weigh them against one another. Where CPFs are identical, then you can bring other factors into play in your decision (aesthetics of the box, say).

Cost Per Unit Function: the cost of each function *that is valued as important to you* that a machine at a given price is capable of handling.

If you want more help in analysis and selection, see the references in Appendix III for doing hardcore systems analysis, and the following text.

Covvey, H. Dominic, and Neil Harding McAlister. *Computer Choices*. Reading, Mass.: Addison-Wesley, 1982.
The authors present methods for assessing the "cost/benefit" ratio for various plans involving microcomputer systems. Thoroughgoing without being overly pompous, this book is recommended for netweavers who are doing it themselves or letting someone else do it (choosing equipment, software, functions).

You can use any of the popular micros as terminals. Some of these machines will also serve well as host nodes for an electronic community conferencing and/ or bulletin board system, allowing people to call up and access messages or programs. If it's simply a matter of going on-line and seeing what's there, a basic setup can get you started: a modem, a low-cost or used (under $500) micro, and a television set (and your telephone, of course). From there, you can add features and more sophistication as your needs dictate. Who knows? You may not *like* all this telecommunicating stuff, anyway.

There are many books, magazines, and users' groups to choose from if your goal is to learn about the different kinds of machines available. Here are the names of manufacturers to write, just in case you're in a location that is scarce on micro info.

Osborne Computer
26538 Danti Court
Hayward, CA 94545
415–887–8080

Tandy Radio Shack
1600 One Tandy Center
Fort Worth, TX 76102
800–433–5502
817–390–3885

Commodore Computer
  Systems, Inc.
1200 Wilson Drive
West Chester, PA 19380
215–436–4200

Apple Computer
20525 Mariani Avenue
Cupertino, CA 95014
408–996–1010
800–538–9696
800–662–9238 (Calif.)

Texas Instruments
Customer Relations Department
P.O. Box 53
Lubbock, TX 79408
806–796–3210
800–858–4565

IBM Corp.
Old Orchard Road
Armonk, NY 10504
800–631–5582
800–352–4960 (N.J.)

Atari, Inc.
1265 Orregas
P.O. Box 3427
Sunnyvale, CA 94088
408–942–6500

Kaypro
533 Stevens Avenue
Solana Beach, CA 92075
619–755–1134

One unique solution to the terminal problem, assuming that you don't want a fully equipped micro at your remote softspace, is a *printer-terminal* combination. The Whisper Writer series from 3M Company is worth checking out if all you need is a way of capturing and printing out data and sending short written replies out. More information from:

3M
Business Communication Products Division
3M Center
St. Paul, MN 55144

4.13
CABLES

The simplest connection is a dyad: two micros at the same location linked together for the purpose of file exchange or conversion between them. Call this the "smallest LAN." The micros must be linked using the proper connectors and with due attention to matching signals.

Plugs and cable are available at your local Radio Shack or other favorite electronics supply store. For making your own RS-232C cable, you will want D-Subminiature flat cable 25-pin connectors (female and/or male) and a length of cable to go with it.

Electronics specialty stores or local computer outlets usually can wire a cable for you quickly according to your specs if doing it yourself boggles you.

4.13.1
RS-232C

RS-232C Pin Signal Designations

Pin	Description	Shorthand
1	Protective ground	
2	Transmit(ted) data	
3	Receive(d) data	
4	Request to send	RTS
5	Clear to send	CLS
6	Data set ready	DSR
7	Signal ground (common return)	
8	Carrier detect (Received Line Signal Detect)	
9, 10, 11	Not connected	
12	Secondary carrier on (Secondary Received Line Signal Detector)	
13	Secondary clear to send	
14	Secondary transmit(ted) data	
15	Transmission signal timing	
16	Secondary received data	
17	Receiver signal timing	
18	Not connected	
19	Secondary request to send	
20	Data terminal ready	DTR
21	Signal quality detector	
22	Ring indicate	
23	Data (signal) rate select(or)	
24	Transmit signal timing	
25	Not connected	

The specification does not determine the physical properties of the connector that plugs into the machine, its dimensions, pin placement, etc. The connector conventionally used is called a DB25. It comes in male and female varieties. The RS-232C spec assumes that a terminal device (DTE) is connected to a modem device (DCE). So when you are linking your two micros, you pick one to be the "modem" and the other to be the "terminal." What follows should be based on that arbitrary decision.

Usually, male connectors are on both ends of the cable. The devices you'll be connecting will usually have female connectors on them. You may from time to time have to connect equipment that varies from this norm, which means making or buying a cable with the appropriate sex on either end. Maleness and femaleness of the connectors will not affect how the cable is actually wired.

In direct, micro-to-micro connections, the only pins you'll have to be concerned with are the following:

Protective ground	Data set ready
Transmit data	Signal ground
Receive data	Carrier detect
Request to send	Data terminal ready
Clear to send	

While you might think that's enough variables to handle, there's more to consider. The RS232C ports, while they look physically the same with their DB25 connectors, come in at least *two* different varieties. We'll call the modem side the DCE (for Data Circuit Equipment) side. The other micro will be the terminal side, or DTE (for Data Terminal Equipment). On the DCE side (the part that plugs into the modem, remember), data is transmitted on pin 2 and received on pin 3.

On the DTE side, data is sent out on pin 3 and received on pin 2. On both sides, the signal is sent as a voltage relative to the signal ground pin 7.

If you plan to do a lot of serious netweaving, plan also to get involved with solder and soldering irons. If your application is one-shot, then you can usually hire someone to wire up the cable for you. But be sure to check the specs of your device manuals in order to tell the person doing the wiring how to wire the pins. Even then you may not be absolutely safe because—gasp!—*manuals can be wrong, too.*

The two micros you will be connecting will probably both be wired as DTE devices at their serial ports. Hence the need for a null modem or crossover cable. This cable will trick the little dummies into speaking to each other. The wire from pin 2 at one end is connected to pin 3 at the other. (See DIAGRAM.)

You can check to see how your particular set of micros is wired by looking in the manuals. If that proves a dead end, or if you suspect that the manual(s) err, then you have to use other means to figure out whether the micro is wired as a DTE or DCE device. This will require using a program that will send data out through the serial port in question. The output pin (whichever it is) will change voltage (measured relative to pin 7) as data goes out. The input pin, of course, won't show any changes at all. You don't have a voltmeter? Then borrow one. Again, serious netweaving, not to mention printerfacing and other electronic feats, will eventually require that you become familiar with the rudiments of a simple voltage meter (available at hobby stores quite inexpensively). If you're a Sunday netweaver, coax a tech-weenie friend into lending you *her* voltmeter.

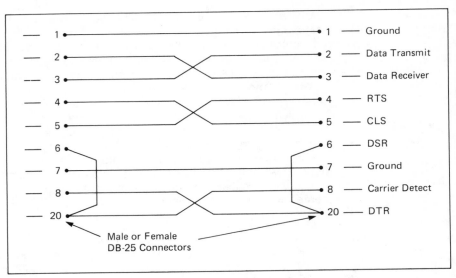

Null Modem Cable Wiring a.k.a. "Crossover Cable"

On some micros—very few—pin 7 will not be connected, despite what the standard says. The maverick manufacturers of these micros, for reasons of their own ranging from ineptitude to laziness, have decided to use pin 1 as the reference or ground signal.

CAUTION: If you run across such a nonconformist micro using pin 1 as ground, check (using a neon tester, another piece of inexpensive indispensable netweaving equipment) to make sure that the builder has not also used pin 1 as the safety ground, connecting it to the chassis of the machine in question. If this is the case, there will be power line voltage on pin 1, and connecting this to pin 7 of another machine can mean *disaster*. Proceed carefully, and if you are not sure, don't do it!

In 99 percent of the cases, the two micros will be ready to talk with just the data lines and signal ground connected. In the remaining cases, handshaking [4.9.1] must occur, which means other wires must also be connected. This is also called *hardware handshaking*. To repeat, most micros will only require the two data lines and the ground to allow exchange.

Two signals at the terminal side are used to tell the modem side what the status of the terminal side is: pin 20, DTR (Data Terminal Ready), tells the modem that the terminal is electrically ready to send, and pin 4, RTS (Request to Send), that it has info and is requesting to send it.

On the modem side, pin 6, DSR (Data Set Ready), reports that the modem is electrically ready. Pin 8, DCD (Data Carrier Detect, or sometimes just CD for Carrier Detect), says that a connection has been made to another modem (perhaps off premises). Pin 5, CTS (Clear to Send), indicates that the modem is ready to accept data from the terminal.

Micros are generally wired, I repeat, like DTE devices. So, if one or more of these status lines are required, you'll have to compensate via your cable wiring. Usually, the only status signal required will be the one saying that the modem is ready (DSR).

If one or both micros need this kind of hardware handshaking, the simplest way to handle it is to wire pin 4 (RTS) directly to pins 5, 6, and 8. If pin 4 does not provide a signal when data is available, then try using pin 20 instead. The latter gives a positive signal whenever the system is powered up.

If both computers are in the same room and both have modems attached, then an even simpler method is to simply connect the two modems directly to each other and fire away. You must, in this case, decide beforehand which micro will be the "originator" and which will "answer."

You can get more assistance concerning the making of fancier crossover cables from *Data Communications for Microcomputers* [Appendix IV: 4.5].

Alternatively, you can write for a copy of the *Black Box Catalog,* which is devoted to devices for data communications and computers, including cables, switch boxes, connectors, and so on. They also supply complete instructions on do-it-yourself cablemaking, a simple process as long as you have the proper tools and some patience.

Black Box Catalog
P.O. Box 12800
Pittsburgh, PA 15241
412–746–5530 (ordering info)
412–746–5565 (technical support: Very Friendly Netweavers)

In the "Why don't they think up more like this one?" category, we have, in this corner, the phenomenal SmartCable from IQ Technologies, Inc. This intelligent RS232C (with DB25 connectors) cable samples the signals of whatever it's connected to, then correctly links the two. The instructions for its use are ridiculously simple, and in tests with both modems and printers, it works flawlessly. This tool is worth having two or three of. It can be used in fieldwork to test two devices to see if they work together before wiring up a permanent cable. My only criticism of the device is that more information should be supplied with its documentation—such as it is—so that the cable could be used for diagnostics in the field. More information on how to order or the name of a nearby dealer from:

IQ Technologies, Inc.
11811 N.E. First Street
Bellevue, WA 98005
800–227–6703
800–632–7979 (Calif.)
800–663–9767 (Canada)

You can obtain copies of the various RS (Recommended Standard) documents by writing to:

Electronic Industries Association
Engineering Department
2001 Eye Street, N.W.
Washington, DC 20006

There will be a nominal charge for copies of the standards you order.

**4.13.3
CENTRONICS**

Copies of the Centronics standard can be obtained by writing:

Centronics
One Wall Street
Hudson, NH 03051

**4.14
SOFTWARE:
PROGRAMMING
AND NETWARE**

How do you find good software of any kind, much less good communications software? Well, here's what you *shouldn't* rely on for evaluating software:

Descriptions in manuals and manufacturers' literature

Catalogs

Books that tell you how to buy software

Magazine reviews

Users'-group recommendations

Dealers' recommendations

Consultants' recommendations

On-line gossip

The three rules for evaluating software are:

1. Use it.
2. Use it.
3. Use it.

The most that any of the secondary information sources can help you do is determine what's out there, and even then you probably won't be seeing all the options. Just as you wouldn't buy stereo speakers or other components without listening to them (would you?), you shouldn't walk out of a store with a piece of software without trying it first, either at the store or somewhere else. Buying by mail order is even less reliable, unless you are finally sure about what you are getting.

Now, you wonder, how can I go ahead and make the specific recommendations that appear in the following sections? Easy. All of the programs listed have been found "tried and true" by a large number of people over a significant period of time. These are pointers: the information is provided as a *starting point* and base line in your netweaving research. Doubtless you will form opinions of your own as you use some of these programs or find others that you like better. (If you do either one, let us know using the feedback form at the back of this sourcebook.)

You can continue your research—an ongoing process in any case—by using a combination of the methods listed above. A good source of information on all kinds of software is:

ITM
Software Division
936 Dewing Avenue, Suite E
Lafayette, CA 94549
800–334–3404
415–284–7540 (Calif.)

ITM is a distributor that runs a search service and has a catalog of software.

4.14.1
INTEGRATED
NETWARE
SYSTEMS

This area in software development is too recent to be "tried and true," but experience is developing fast. Some of the integrated systems use a disk file format called DIF (Data Interchange Format), which enables many different kinds of programs on differently formatted disks to share the same data. It also makes shipping data around on the phone lines easier between these families of programs. For more information on the DIF family approach, see the November 1981 *Byte* or write:

DIF Clearinghouse
P.O. Box 638
Newton Lower Falls, MA 02162

4.14.2
TERMINAL
NETWARE

PC-Talk III is a good fundamental terminal package for the IBM PC. It is low-cost, does the same thing as packages many times its price, and works consistently well according to most users and reviewers. It uses the XMODEM protocol. Write for current price and availability information:

Headlands Press
P.O. Box 862
Tiburon, CA 94920

Telestar is a terminal program for the Radio Shack Model 100 portable. It is designed for on-the-road use. 606–739–6088. (A BBS after 6:00 P.M.)

4.14.3
SMARTER
NETWARE

What makes a terminal program "smart"? Here's a list of the basic features to look for, ranked more or less top to bottom from "necessary" to "nice."

1. Does the program support both pulse dialing and tone dialing? Tone dialing is necessary to access some of the alternative phone services.

2. Can the program "capture" text by setting aside a buffer area in the

micro's memory? How big is the buffer? Will the size of the buffer handle the kinds of messages and text material you'll be dealing with?

3. Does the program use a buffer to transmit text as well? If so, can you prepare your text off-line and save it on disk for later transmission? Is the text editor provided easy to use? Will it transmit multiple files during the same on-line session without making you disconnect and redial before transmitting the next file? (Believe it or not, some programs require this, and they are clumsy, to say the least.)

4. Does the program make it easy for you to turn on your printer while on-line, so that you can print out material?

5. Does the program give you easy access to all the protocol parameters while you're on line? (baud rate, duplex/half duplex, data bits, parity, stop bit).

6. Does the program have a dialing directory and a means of creating keyboard macros, so that you can access and log on to a given service with just a few keystrokes? Does it support autodialing? Automatic redialing of busy numbers?

7. Does the program support your particular modem? Will it support 1,200 baud without "dropping characters"? This essentially means that the software is written to be fast enough to keep up with the modem and your micro at the same time.

8. Does the program support both the XMODEM (Christensen) and X-on/ X-off protocols?

9. Can the software generate a true BREAK signal (the transmission line is held at 0 for a minimum time period, between 100 and 600 microseconds)? This is only needed, of course, if you are planning to connect with computers that can respond to this signal, such as some office mainframes. For most on-line work, you won't need it anyway.

10. Does the program provide for translation tables? If so, how big can the tables be? Five characters? Ten? Twenty?

11. Can you transmit both machine language and text files? This is important only if you plan to do a lot of program swapping on line. For most conferencing and netweaving, you won't care.

12. Does the program provide for *upload throttling,* in which extra time is put in at the ends of lines of transmitted text? This feature may be necessary when connecting to some machines. Just another extra thing to check out.

With these criteria in mind, I have found the following terminal programs to be the best in their categories, for the following micros.

Apple II, II+, and IIe.   For regular Apples, *ASCII Express,* and *Pro ASCII Express,* from Southwestern Data Systems, have been out for more than two years, are thoroughly debugged, and work wonderfully. "The Professional" package costs more than *ASCII Express,* but it does a whole lot more, and studying it thoroughly is a complete communications course all its own. For Apples using one of the CP/M cards, this same company provides ZTerm and Pro ZTerm. Again, from personal use both of these packages are excellent and well tested. Source:

United Software Industries
1880 Century Park East, Suite 311
Los Angeles, CA 90067
213–556–2211

It should also be noted here that many of the medium- to high-priced modems come with their own terminal software, which varies in quality from manufacturer to manufacturer. However, these packages may be enough for you to get by with.

TRS-80 Model 100. For on-the-road work, nothing yet comes close to the TRS-80 Model 100 "lap micro" for ease of use and functionality. It can be used as a remote access terminal and comes with a built-in modem and communications programs. If mobility is important to you in telecommunications, this is the one.

TRS 80 Color Computer. There are several quite good terminal programs for the CoCo, as its fans refer to it. One is *Autoterm*.

PXE Computing
11 Vicksburn Lane
Richardson, TX 75080
214–699–7273

In order to use both a printer and a modem at the same time with a CoCo, you need to get an adaptor called the Microtext Communications Module. This module plugs into the cartridge slot in the CoCo. It also contains terminal software in its ROM. From:

MicroWorks
P.O. Box 110
Del Mar, CA 92014

Commodore. The VIC 20 is an excellent low-cost way to begin in the world of telecommunications. Although these are no longer being made, there are still a lot of them available, and used ones are already appearing in the marketplace. Midwest Micro Associates makes two fine terminal programs for the VIC 20: *Terminal-40* and *SuperTerm-40*.

Midwest Micro Associates
P.O. Box 6148
Kansas City, MO 64110

The same group is a good source for Commodore 64 terminal software: *64 Terminal* and *SuperTerm 64*.

Atari. *Teletari* requires at least 32K of memory in your Atari, with at least one disk drive. It is a good basic terminal program. From:

> Don't Ask Software
> 2265 Westwood Blvd.
> Suite B-150
> Los Angeles, CA 90064

Teletalk is also a good program available for Ataris. It is written in machine language, which makes it a tad faster than *Teletari*. From:

> Datasoft, Inc.
> 19519 Business Center Drive
> Northridge, CA 91324

Finally, *T.H.E. Smart Terminal* is a proven telecommunications program for your Atari and can be recommended safely. From:

> Binary Computer Software
> 3237 Woodward Avenue
> Berkely, MI 48072
> 313–458–0533

**4.14.4
MICRO-TO-
MAINFRAME
NETWARE**

In order to connect a micro to some mainframes, you have to make the micro appear to be the kind of terminal the mainframe is accustomed to dealing with. This can sometimes be done in software, using translation tables and the like. More often, another piece of hardware is required, either inside the micro or between the micro and the mainframe.

Putting an IBM PC micro into the bisynch mode (IBM synchronous) so that it can talk with its own kind in the IBM mainframe world is made easy with AST 3780, an IBM 2780/3780 emulation package. Source:

> AST Research, Inc.
> 2372 Morse Avenue
> Irvine, CA 92714
> 714–540–1333

IBM PC to IBM 3274 or 3276 Cluster Controllers (SNA or Bisync). Coax/3278: This interface card for the IBM PC or XT provides a high-speed coaxial port so that it can be (locally) attached to IBM's SNA network and thus access the supporting mainframe. This turns the PC or XT into a "3278 display device." A terminal for the mainframe, by any other jargon. Source:

> Personal Systems Technology, Inc.
> 15801 Rockfield Blvd., Suite A
> Irvine, CA 92714
> 714–859–8871

Coax Link: Another local link turning the PC or XT into a 3278 display. Source:

Micro Link Corp.
P.O. Box 113
1850 W. Wayzata Blvd.
Long Lake, MN 55356

The hardware solution to connecting micros with mainframes is sometimes called "front-end processing." One company making moderately expensive ($1,000+) front-end processing equipment for micros is:

Winterhalter, Inc.
P.O. Box 2180
Ann Arbor, MI 48106
313–662–2002

Their micro product is called *Datatalker*.

4.14.5
E-MAIL AND
CBMS NETWARE

TeleCOMM, an e-mail program for Televideo, CP/M, MP/M, TurboDOS micros. This program provides password protection on files; it supports the following features:

broadcasting and forwarding	written in assembly language
menu-driven user interface	supports popular modems
formattable printouts	send text messages *or* programs
save and delete messages	incoming mail prompt

It's available from:

International Computers and Telecommunications, Inc.
932 Hungerford Drive
Rockville, MD 20850
301–251–0062

Expensive solutions to the electronic messaging problem are offered by the following firms. Needless to say, they have to be cost-justified but are generally reliable once implemented.

Comet, a service of	Dialcom (a division of ITT)
Computer Corporation of America	1109 Spring Street, Suite 410
675 Massachusetts Avenue	Silver Springs, MD 20910
Cambridge, MA 02139	301–588–1572
617–492–8860	

Both major packet networks provide e-mail services as part of their offering:

Telemail	OnTyme II
GTE Telenet	Tymnet, Inc.
8229 Boone Blvd.	20655 Valley Green Drive
Vienna, VA 22180	Cupertino, CA 95014
703–442–1000	408–446–7000

In addition, some of the major alternative telephone services, such as MCI, are beginning to provide their own e-mail services to subscribers. Some are even adding gateways and black boxes to allow you to send mail to someone who subscribes to a different service, such as Compuserve.

Uncle Sam is also offering a sort of hybrid electronic-to-print e-mail. Called E-Com, the system lets you transmit your letter to the post office, where it gets transmitted to the city of the addressee, gets printed out, stuffed into envelopes, and mailed for a total of two days' delivery time. Messages are limited to two printed pages. Cost: about twenty-six cents for the first page, five cents for the second page, plus attendant costs (communications network fees, etc.). For more information, contact your local postmaster.

Western Union: Telex, TWX, and Mailgrams.   Telex (telegraphic exchange) and TWX (teletype exchange) are telegraph-based communications services. Both constitute separate networks owned and maintained by Western Union, using the regular telephone lines for interconnection. Business and organizations throughout the world use these text-oriented machines to get messages to one another quickly. Your micro softspace can extend to any one of more than a million TWX and telex machines.

Bear in mind that telex and TWX technologies are historically ancient by electronics standards. The telex machine dates back to the 1930s, and the TWX was developed in the 1940s. Both are slow. The sending and receiving stations are electromechanical. You can think of them as communicating printers or hardcopy phone answering machines.

Using an intermediary service provided by Graphnet, you can connect your personal or small business softspace to any of the installed telex or TWX machines in the world using a micro with appropriate communications software. The converse is also possible: your micro can be the receiving station for telex and TWX messages from existing installed machines (or micros, for that matter, also connected through Graphnet's services).

A Mailgram is transmitted from one Western Union office to another. At the destination, a hardcopy of the message is printed out and put into the local U.S. mail for local delivery. Mailgrams are usually delivered overnight, and have as their main advantage over regular telegraph that a hardcopy results. This service can also be accessed with your micro.

Telex/TWX/Mailgram Services for Micros.   Graphnet is a mainframe-based service that acts as a gateway [4.21] to Western Union's network. You can send messages to any telex or TWX machine, or send a Mailgram through Graphnet.

Message traffic has to be pretty high to justify subscribing to Graphnet, but it may just fit the needs of some users who do a lot of communicating with the business community.

Graphnet, Inc.	or	Graphnet Telemarketing
329 Alfred Avenue		Department
Teaneck, NJ 07666		8230 Boone Blvd., Suite 330
800–631–1581		Vienna, VA 22180
800–932–0848 (N.J.)		800–336–3729
		703–556–9397 (Va.)

Many word processing software packages can be used to prepare your messages before sending them to a remote computer. However, some of these packages, most notably WordStar, but some others also, create nonstandard text files that must be "tweaked" or fixed before you can send them. Check with your dealer or supplier of your particular W/P package to make sure that you can prepare documents for uploading without difficulty. There are a *few* so-called communicating word processors on the market. One for the Apple II, II+, and IIe that we've had good experiences with is ZARDAX. More information about this package is available from:

Action-Research Northwest
11442 Marine View Drive S.W.
Seattle, WA 98146

Another service-bureau approach to Western Union's network for micro users is available from:

SpeediTelex International
3400 Peachtree Road N.E.
Atlanta, GA 30320
800–241–1913

Chat II (hardware) is a "black box" solution for accessing Telex and TWX networks, based on an 8085 microprocessor. Can also send Mailgrams. Source:

Chat Corporation
2560 Wyandotte Street
Mountain View, CA 94043
415–962–9670

MICRO/Telegram is a "black box" in software to enable your micro to access Telex and TWX networks. From:

Microcom, Inc.
1400A Providence Highway
Norwood, MA 02062
617–762–9310

MicroTLX is a CP/M-based software solution to link to Western Union networks. Source:

Advanced Micro Techniques
1291 E. Hillside Blvd., Suite 209
Foster City, CA 94404
415–349–9336

A note on telex. The telex network is a half-duplex system. Therefore, you must transmit your data in the half-duplex communications mode. When accessing most of the modern networks, such as TYMNET or Telenet, when you send out a CR (carriage return), you'll get a LF (linefeed) in return. Not with telex. When you send out a CR, all you get back in return will be a CR. This means, practically speaking, that you'll have to make sure that you have put your own line feeds in the copy you send, or else the remote machine to which you are transmitting will overprint lines, resulting in garbage at the other end. Sigh. In any event, telex-specific software should take care of these matters.

4.14.6
BBS NETWARE

Just about any micro can be a host computer for an on-line network. The kinds of things that can be done with such systems are quite far-ranging. Based on experience, though, you shouldn't start out with any less than the following:

- At least 48K of RAM in your host computer system
- At least two disk drives
- A dedicated incoming-call phone line
- A printer
- 1,200-baud support

Anyone can start a community bulletin board or conferencing system. But it takes some work to keep it going and keep it interesting. It also entails some responsibility. For the popular CP/M-based systems, you'll need some working knowledge of CP/M and, helpfully, some Z80 assembly language know-how, too. Such knowledge isn't necessary with some of the other systems, such as the Apple BBSs or Conference Trees (see following).

The CP/M systems come in two distinct "flavors." The RBBS flavor is just a one-shot message system: users call up, read messages, and enter messages. The RCPM flavor allows users to get into the host computer's CP/M operating system, which, in turn, allows the transfer of program files as well as text. The BYE program [see Appendix IV: 4.15] is a "remote console" program that is the guts of CP/M BBSs. The public domain RBBSs are the most popular, right now. Version 3.1, written in Microsoft Basic (MBASIC 5.2), is a good one. It has to be compiled using Microsoft's BASCOM compiler, but this is a pretty straight-forward process. Early 1984 cost: $300–$400.

File transfer is accomplished using one of the MODEM terminal programs, then using the remote system's XMODEM program to do the actual file transfer. Most CP/M files are saved on the host system's disk in a "squeezed" format, to allow faster transfer (especially at 300 baud). A squeezed file is a regular CP/M file that has been condensed using SQ-15COM, another public domain program. Remote CP/M systems are much more vulnerable to those wishing to damage or rummage.

For the Apple II, II +, and IIe, the best general purpose bulletin board system is *NetWorks II*.

An economical way to begin a plain vanilla electronic bulletin board system is with a TRS-80 Color Computer (CoCo). *Color-80* is BBS netware from:

Silicon Rainbow Products
663 S. Bernando Avenue
Suite 225
Sunnyvale, CA 94087

**4.14.7
CONFERENCING
NETWARE**

Any communications medium slants the contents of its messages in certain ways. This means that all communications media have *biases* that will tend to exclude some forms of communication and include others. New modes mean new possible kinds of communication, and new constraints. Good *conferencing* software should encourage just that: valuable information exchanges in the context of an electronic conversation with others. Bulletin board, RBBS, and similar kinds of computerist-oriented systems encourage computerist sorts of exchanges: gossip about and for infopiracy, pro and con, computer programs and program fixes, and the regular chitchat. Conferencing systems, on the other hand, lend themselves to on-line soap opera simulations, with the writers being the role players being the readers, each person in the network assuming a character or persona and developing imaginary on-line plots. Or they become living electronic journals, with ideas in development having the advantage of peer review and comment *before* regular print publication. Or they mutate into yentas, finding romance for the lovelorn at all hours of the day or night. In short, good computer conferences come in as wide a variety as good professional conferences (face to face) of all kinds, and are least constrained by the micro-computer setting.

Conferencing Software for the IBM PC, TRS-80, and Apple II, II+, and IIe. The conference tree concept was pioneered by the CommuniTree Group in San Francisco. The concept has since been adopted by at least one mass information videotext service. Its name comes from the structure of its database, which looks like branches off branches off a main "trunk." Any new branch can grow into a whole tree, and anyone can add a new branch to any existing branch. This makes lines of thought in on-line conversations very easy to follow and to organize after the fact, for possible hardcopy printing.

The CommuniTree software works essentially the same no matter which micro is the host system. No special communications packages or proprietary protocols are needed to access conference trees from remote locations. Information on price and availability from:

CommuniTree Group
1150 Bryant Street
San Francisco, CA 94103

**4.15
NETWEAVING:
GOING ON-LINE**

Although librarians and archivists love them, netweavers should be selective in their purchase and use of directories. The information nearly always overlaps in any two you select, and at least two should be on hand rather than just one. The information is nearly always partially obsolete by the time you get even a current copy. And finally the information is never "comprehensive," no matter what the claims of the particular directory publisher may be. Directories are only as useful as their organizing strategy and ease of use. Mere lists or arbitrary groupings are next to useless. Costly directories and newsletter services are not necessarily better than less expensive ones. Frequent updating is valuable only if you use a reference frequently. The best directories *should* be on-line. So far, most are not, and those that are, are too expensive.

The Directory of Online Databases is one of the oldest and best of such directories. It is updated frequently. It is also one of the most expensive. Many of the data bases listed are carried by other services, such as Dialog.

> *Directory of Online Databases*
> Cuadra Associates, Inc.
> 2001 Wilshire Blvd., Suite 305
> Santa Monica, CA 90403
> 213–829–9972

Although the term "information industry" is pretty broad, covering software and book publishers, database providers and hardware manufacturers, *The Information Industry Marketplace* is a useful pointer source to have in a net-weaver's reference library. R. R. Bowker, the publisher, is an old hand at information, having published guides to the publishing industry for a long time. The organization and format of Bowker directories is useful and suggestive. Reasonably priced and worth every penny.

> *Information Industry Marketplace*
> R. R. Bowker & Co.
> 1180 Avenue of the Americas
> New York, NY 10036
> 212–764–5100

Another useful guide is the:

> *Encyclopedia of Information Systems and Services*
> Gale Research Company
> Book Tower
> Detroit, MI 48226

Yes, but what *is* there to go on-line with? Alas, fair netweaver, the world is vast and maps are few. The above directories will refer you to some of the following sources as well. In any case, here are some places to begin.

At the local level, there are small-scale, dedicated systems of all kinds. These are called, variously, Bulletin Board Systems (BBS), Community Bulletin Board Systems (CBBS), conference trees (C-Trees), Remote CP/M systems or remote bulletin boards (RCPM and RBBS), "underground bulletin boards"—why "underground" I can't fathom—and community networks. These are all versions of SPINs: special-interest networks. They all run on smaller machines, generally support only one telephone line—hence user—at a time, and each has its advantages and disadvantages.

For listings of these kinds of systems, there are three good places to begin.

Numbers can be found on-line, placed there by users or creators of other SPINs. There is one premier source of such on-line listings of other on-line numbers:

PMS-Santee
619–561–7277 (modem)
300 baud
even parity
7 data bits
1 stop bit

Or by mail at:

Datel Systems
Box 817
Lakeside, CA 92040
619–443–6616 (voice only, please)

The *User's Guide to CP/M Systems and Software,* from TUG, Inc., contains complete listings of RCPM systems, and is oriented toward CP/M micro (Z-80) users. Write for price and availability info:

TUG, Inc.
Box 3050
Stanford, CA 94304

For a listing of C-tree systems, send a stamped, self-addressed envelope to:

Conference Trees
CommuniTree Group
1150 Bryant Street
San Francisco, CA 94103

One of the better printed sources of SPIN information comes from:

On Line Computer Telephone Directory
P.O. Box 10005
Kansas City, MO 64111

Write for price and availability information. The numbers listed are said to be tested for accuracy using a computer to dial up and verify each number before it can be listed in the OLCTD.

Of the mass-based videotext services, we have found the most usefulness in two, but that's us, too. Our uses at CommuniTree Group have been oriented toward research in our own areas of special interest. We find bibliographic databases, while being somewhat more expensive per use than the mass-based services, give more infobang for the buck: denser, more relevant, more current. Also, since they don't have such things as electronic mail or bulletin boards or conferencing, research databases save *that* source of on-line budget squeeze. (Besides, it's easier, more fun, and less costly to do most bulletin board hopping on the local level, using free access numbers and the public phone lines.) Knowledge Index, from Dialog, is a good buy, overall.

Knowledge Index
Dialog Information Services, Inc.
Marketing Department
3460 Hillview Avenue
Palo Alto, CA 94304
800–528–6050
800–352–0458 (Ariz.)   } all ext. 415
800–528–0470 (Alaska and Hawaii)
Available through Tymnet and Telenet

A second, up-and-coming service is BRS After Dark, another bibliographic reference service. BRS, however, plans to offer electronic mail and a bulletin board service for users as well.

BRS After Dark
1200 Route 7
Latham, NY 12110
518–783–1161

The three most heavily marketed, if not the most popular and frequently used, of the mass-based services are the Source, Delphi, and Compuserve. The best source of information about all three is the internal marketing departments of the companies. Unless you enjoy throwing your money away on otherwise free information, you'd be ill-advised to spend bucks on any books shuckin' and jivin' you through an "imaginary" session on any of these services: all of them give you some free time to determine for yourself whether you like them or not.

Our ranking, best to worst, of the mass services is as follows:

Compuserve Information Service
5000 Arlington Centre Boulevard
Columbus, OH 43220
614–457–8600

Delphi
General Videotex Corporation
3 Blackstone Street
Cambridge, MA 02139
617–491–3393
800–544–4005

The Source
Source Telecomputing Corp.
1616 Anderson Road
McLean, VA 22102
800–336–3366
800–572–2070 (Va.)
703–734–7500 (Outside U.S.)

Public Domain Software.   Software that you can obtain for the cost of the search alone, or for the cost of a disk and some postage, is called public domain software. This "freeware" will require that you take time, disk space, and sometimes phone charges to get this stuff, and get it working right in your machine. In most cases, such software resides on diskettes in user-group libraries. A similar and overlapping collection is also available on micro networks, via the grapevine of bulletin board, network, and conferencing systems.

The best public domain software is oriented toward professional or serious amateur programmers. They include BASIC and Z80 (CP/M) program subroutines; utilities such as file and/or disk management; and programming "oddities" that teach some algorithmic truth in striking and original ways. It's as though the chefs of the world were to get together and swap portions or all of their best recipes.

Many of these recipes are too small, or too limited in market potential, to bear the cost of packaging, separate written (much less printed) documentation, or (horror of horrors) product support. They go into the world of public domain to sink or swim on their own. By now there is reason to believe, based on collective experience with them, that some will die out and others will survive.

If your softspace micro uses or supports CP/M, then you have a treat in store for you. Many Remote CP/M (RCPM) bulletin board systems have libraries of software available for downloading and use in your micro. CP/M based software is probably the largest collection of public domain software.

Some representative COMMUNICATIONS PD packages available:

MDM712 (a MODEM7 update), MODEM75, and COMM—terminal/communications programs

BYE and RBBS—BBS programs for running host systems

XMODEM—downloading/uploading terminal program

USQ—decompressor for programs using XMODEM

CHAT—real-time talky-talky between two micros

HOW2SEND and PROTOCOL—documentation on communications protocols (including modem-xmodem protocol)

Sources for CP/M public domain software:

- PicoNet RBBS at: 415–965–4097 (modem number).
- You can find most of the BBS systems in the country listed on People's Message System BBS in Santee, California: 619–561–7277 (modem).
- Also, PCMODEM BBS at 312–259–8086 (modem), has listings of other BBSs and small-scale community conferencing systems.

- P-Chicago at 312–944–4849 (modem), has the popular PCTalk III, for the IBM PC.
- A good source of disk-based PD freeware is:

  PC/Blue Library
  c/o The New York Amateur Computer Club
  P.O. Box 106
  Church Street Station
  New York, NY 10008
  212–864–4595 (voice hotline)

  Send them an S.A.S.E. for an updated list of PD Disk volumes.

This group distributes a program called KERMIT, which is a mainframe to IBM PC communications program (ask for vols. 27 and 28). KERMIT supports VT52 emulation. Baud rates of 300–9,600 are available. Also supports IBM's VM/CMS o.s. and DEC 20 mode. Columbia University holds the copyright on the program, but it can be freely exchanged as long as it is not sold.

Some of these public domain memes evolve, that is, get improved by other programmers who pick them up. The telecomm program called MODEM (MODEM7 a.k.a. MODEM75) is an example of a program that has improved through use. Written by Ward Christensen in 1979, it has gone from a simple file-transfer program to a sophisticated infosculpting tool that uses CRC (Cyclic Redundancy Checking: see Glossary) error detection. In the view of many, MODEM is now a better program than any of its type that you would have to pay for commercially.

Caveat: Some commercial sources are *selling* public domain software at what-the-market-will-bear prices: the same software that is available free on the BBS's or in user-group libraries.

Another communications program useful to netweavers is a public domain package called BYE, also written by the legendary Mr. Christensen. BYE allows the PC to act as a host BBS, with uploading, downloading, and messaging features. A co-program called Sweep 38 copies, edits, and moves files created by the BYE BBS program. XMODEM 74 embodies the Christensen communications protocol as well.

A public domain modem program called MBOOT supports your modem. It is written is Z80 assembly language and is available in source form (printed) for your CP/M micro. See the New York Amateur Computer Club catalog, above.

About 1 percent of the programs in the public domain are worth having locally, for use in your own softspace. The rest are either too esoteric or too awful to bother with.

Other public domain programs that have been identified as better than average include the following.

ZCPR for the Z80 replaces the CCP (Console Command Processor) in the operating system and adds features that make CP/M much easier to use. It is especially useful to netweavers because it prevents unauthorized users from altering (by erasing or renaming) files.

The TRS-100 lap micro has a special-interest group on CompuServe that serves up free software for this machine. Type GO PCS-154 from any ! CompuServe prompt. Here you can also find RBBS-PC, another program for the creation of a PC-based BBS, this one written by Russ Lane and modified by Brad Hanson and Larry Jordan.

Some estimates have put the number of public domain programs available at 15,000+. At least two books on the subject are in preparation, as of this writing, so that at least some preliminary sifting work will be done for you soon. After eliminating redundant or dual-named identical programs or programs that are only minimally different from one another, eliminating the esoteric and the just plain junk, and then getting rid of utilities that are trivial, one *would* probably be left with 100–150 useful programs. Of these, some will be on disk, in users' libraries, and some will be on the networks, on BBSs and user-group sections on the Source and Compuserve. Some may also be in print.

4.16
LOCAL AREA
NETWORKS (LANs)

Purists who brandish authentic telecommunications credentials may quibble, but netweavers may define *any* sharing of peripherals or connecting of two or more micros as a Local Area Network. One of the simplest ways to accomplish this is to install a switchbox that will allow two computers to access the same printer. The switch box assures that you won't have to run around changing cables everytime you want to send something to your printer. Some writers I know use such a scheme to alternate two printers on a single computer: a dot matrix printer for draft work, a letter-quality printer for final manuscripts.

In addition to the Black Box people [Appendix IV: 4.13.1], another company offering quality switches for small local networks:

Advanced Systems Concepts, Inc.
435 North Lake Avenue
Pasadena, CA 91101
818–793–8979

4.17.2
WIRING
TECHNIQUES AND
CONNECTIONS
(LANs)

Personal computers, word processors, and terminals can communicate with one another over existing local PBX (Private Branch Exchange) telephone systems. ComNet 48 is a modem using existing twisted pair wiring (used in most PBXs and PABXs) within a building. This modem does not require a central switch, any additional cabling, or any changes in the main distribution frame (that maze of wires and connections usually located in a company closet somewhere). It plugs right into an existing phone wall jack and is ready to go. Both a terminal and a telephone then can be plugged into the modem. This system allows full-duplex, two-wire synchronous or asynchronous data communications at 4,800 bits/sec. Source:

Avanti Communications Corporation
Aquidneck Industrial Park
Newport, RI 02840
401–849–4660

4.21
BRIDGES AND
GATEWAYS AND
LINKS (OH, MY!)

Microwave Links. General Electric offers a short-haul, off-premises communications link called GEMLINK. This can serve as a bridge between two LANs in two different office buildings. It offers 48 voice channels and/or 240 digital data channels. It operates up to several miles, using the 23 GHz frequency band. It interfaces with a wide variety of equipment at speeds up to 256Kb/s, 1.544Mb/s, and 3.152Mb/s. Intracity short-haul microwave is an alternative to local leased lines, cable, or conventional microwave, all of which tend to be more expensive than GEMLINK. Source:

Microwave Link Operation
General Electric Co.
316 E. Ninth Street
Owensboro, KY 42301

**4.23
SOFTSPACE
EXPANSION AND
SCALING UP**

Creating an integrated softspace, capable of handling voice, text, images (drawings, slides, moving pictures, NAPLPS) in a variety of creative, display, or processing modes should be one of the long-term goals of any netweaver. There is a wide variety of ways to customize your working information environment. For example, did you know you can get a device for your phone that will let you know who is calling *before* you answer the phone? Very useful for anyone with a demanding client network. The device is called Privecode. It is marketed as a "telephone access control terminal" by International Mobile Machines Corporation. With a voice-synthesis chip, this microprocessor-controlled phone accessory asks your callers to enter their own personal three-digit identifying code. The machine then displays the code on the terminal before you answer the phone. Handy, for those who need an extra private secretary for their private phone. Source:

IMM
100 North 20th Street
Philadelphia, PA 19103
800–523–0103, ext. 110
215–569–1300, ext. 110 (Pa.)

Two good sources of high-tech consumer items in telecommunications can be had for the asking. These two companies provide their catalogs filled with micros, telephones, answering machines, and more for the price of a phone call:

JS&A Products That Think
One JS&A Plaza
Northbrook, IL 60062
800–GADGETS

Markline
P.O. Box C-5
Belmont, MA 02178
800–225–8493
800–225–8390 (Mass.)

**4.23.1
THE PAPER
CONNECTION**

Typesetting Companies. The following list of firms is meant to help you get started in this area. Contacting one or more of them may elicit leads to others, in other cities. Soon, as typesetters realize what a fantastic new market for their services this really is, most any firm will be able to help you. Assuming, of course, that we get some semblance of standards in the field.

Amnet
1015 Gayley, Suite 288
Los Angeles, CA 90024
818–907–5015

Pacesetting Services
200J North Crescent Way
Anaheim, CA 92801
714–956–0860

Buckland Printing Company
P.O. Box 1157
99 Willie Street
Lowell, MA 01853
617–452–0111
617–458–2522

Bye and Bye, Inc.
110 South Main
Holstein, IA 51025
712–368–4353

Cimarron Graphics
Box 12593
Dallas, TX 75225
214–691–5092

Typesetting Communications Products.   For TRS-80 and S-100 systems using CP/M there is direct connection to hook micros directly to MicroComposer type-setting equipment. This lets your micro be the "front end" of the composition process through to setting of type. Source:

Cybertext Corporation
P.O. Box 860
Arcata, CA 95521
707–822–7079

Teleset-80 is a typesetting communications package for the TRS-80 (it must be used only with Scripsit word processing files or equivalent). Source:

Small Business Systems Group
6 Carlisle Road
Westford, MA 01886
617–692–3800

Typesetting interface software and information are available for TRS-80s, Apples, IBM PCs and others from:

First Main Computer Systems, Inc.
P.O. Box 795
Bedford, TX 76021
817–540–2491

Print and Micro Symbiosis.   Print will point to computer sources. Computer sources will point to print. Both media will live side by side for quite a long time to come. You can access a lot of information on microcomputers that lives primarily in magazines through electronic sources. The *Microcomputer Index* is a reader's guide to micro periodicals. You can scan it on Dialog's Knowledge Index using your micro and modem.

Microcomputer Index
2464 El Camino Real, Suite 247
Santa Clara, CA 95051

And you can subscribe to on-line electronic newsletters of all kinds (and growing), including high-tech subjects like telecommunications. Contact:

Newsnet
945 Haverford Road
Bryn Mawr, PA 19010
800–345–1301
215–527–8030 (Pa.)

Glossary

This glossary is intended to provide the netweaver with a basic vocabulary reference. Items have been selected on the basis of most common use in telecommunications, networks and the associated concepts found in the first three chapters of this book. There are several dictionaries available that you will also find useful in acquiring more of the necessary jargon of data communications. Among them:

The Random House Dictionary of New Information Technology. Edited by A. J. Meadows, M. Gordon, and A. Singleton. New York: Vintage Books. Paper.

Sippl, Charles J. *Data Communications Dictionary.* New York: Van Nostrand Reinhold Co. Paper.

Weik, Martin H., D.Sc. *Communications Standard Dictionary.* New York: Van Nostrand Reinhold Co.

Note: Words in capitals inside definitions are themselves entries in the glossary.

ACK The name of a control character transmitted by a receiver to acknowledge or affirm that a signal or group of signals such as a block of characters sent by a transmitter has been received.

ANSI American National Standards Institute, the main standards development organization in the U.S. ANSI is a nonprofit organization supported by trade groups, professional associations, and corporations. It is a nongovernmental representative in the ISO (International Standards Organization).

ARQ Automatic Request for Repetition. A character sequence that causes an automatic retransmission of a signal in which an error has been detected (through PARITY CHECKING etc.).

ASCII American Standard Code for Information Interchange. Developed by ANSI, this standard specifies a seven-bit digital code for each of the ninety-six displayable characters on a micro keyboard. It also specifies thirty-two CONTROL CHAR-ACTERS, which are usually not displayable on video terminals or printers. ASCII has also been adopted in Europe. Only seven bits (128 possible combinations) are needed to specify a particular character. An eighth bit is sometimes used for PARITY checking, hence ASCII is sometimes referred to as an eight-bit code.

Access line   The telephone wire CHANNEL that connects the equipment on your premises to your local carrier office. Also called a LOCAL LOOP.

Acoustic-coupled modem   A MODEM that sends and receives signals (specified sound tones) through an ordinary telephone handset. The handset is inserted into two foam cups on the modem. Thus, the connection between the modem and the phone lines uses sound waves as an intermediary signal.

Albert   A machine that combines TELETEXT, TELEX, word processing, and telephone functions in one unit. Coined by the British Telecom people.

Alphanumeric   Designating both alphabetic and numeric information. See also BYTE, ASCII, and CHARACTER.

Analog signal   An electrical signal consisting of a continuous range of voltages (amplitudes) or frequency values. Most telephone voice services use analog signals. See also DIGITAL SIGNAL.

Answer/originate modem   See AUTO-ANSWER MODEM.

Asynchronous transmission (asynchronous communication)   A data communications mode in which the timing of the signal being sent is not critical. In this mode, each transmitted character includes a start bit and may also include one or more stop bits. Thus, accurate reception of the data does not depend on precision timing between the transmitting machine and the receiving machine. See SYNCHRONOUS TRANSMISSION.

Authorization code   A number assigned to customers of dial-in telephone networks that gives them access to use of the network and identifies them for billing purposes. This code is usually four to six digits long and can be tapped out on a Touch Tone telephone keypad or from the terminal keyboard.

Auto-answer modem   A type of DIRECT CONNECT MODEM that can automatically answer an incoming call and establish connection between the local micro and the device making the call (usually another micro or terminal). Auto-answer modems can usually autodial, letting you type a phone number from the keyboard. The computer/modem combo then dials the number.

Bandwidth   The frequency range of a channel. The smaller its bandwidth, the less information can be carried on a given channel.

BASIC   Beginner's All-purpose Symbolic Instruction Code. An early high-level computer language designed to be easy to learn and use by people with no mathematics or computer background. BASIC is, for all practical purposes, universally available on contemporary microcomputers.

Baud/baud rate   In telecommunications, a unit of signaling speed. The baud rate is the number of discrete signal changes per second. If each signal change represents a single bit, then the baud rate is the same as the number of bits per second. However, in some coding schemes, one signal change can represent two or more bits (see DIBIT, TRIBIT, and QUADBIT). In such cases, the baud rate will not exactly equal the number of bits per second transmitted.

Baudot code	The five-bit or five-channel code used by TELEX systems. This code was devised in the nineteenth century by Emile Baudot. In the United Kingdom, the telex code is called the Murray code.
Bell 103	AT&T specification for a MODEM providing ASYNCHRONOUS originate/answer transmission at speeds up to 300 bits per second (300 BAUD).
Bell 201	AT&T specification for a MODEM providing SYNCHRONOUS data transmission at 2,400 bits per second.
Bell 202	AT&T specification for a MODEM providing ASYNCHRONOUS data transmission at speeds up to 1,800 bits per second. Requires four-wire line for FULL-DUPLEX transmission.
Bell 208	AT&T specification for a MODEM providing SYNCHRONOUS data transmission at 4,800 bits per second.
Bell 209	AT&T specification for a MODEM providing SYNCHRONOUS data transmission at 9,600 bits per second.
Bell 212	AT&T specification for a MODEM providing full-duplex asynchronous or synchronous data transmission at speeds up to 1,200 bits per second on the voice (dial-in) phone line network.
Bell-compatible	A term sometimes applied to MODEMs. A modem is said to be "Bell-compatible" if it conforms to the technical specifications set forth by AT&T for the various devices, such as BELL 212.
Bibliographic database	A collection of information, stored in computer-accessible memory, about books, magazines, journals, and other printed documents.
Biocomputer	Organic brains. The Central Processing Unit (CPU) of carbon-based lifeforms. In humans, the brain/mind system.
Bisynch/ bisynchronous transmission (BST)	An IBM communications protocol involving the continuous exchange of synchronizing signals between two telecommunicating devices. A defined set of CONTROL CHARACTERs is used for the synchronization.
Bit	A contraction of "binary dig*it*." The two binary digits are 0 and 1. Micros generally work only with binary numbers expressed in strings of bits. A BYTE is eight bits. A SLICE is two bits, and a NYBBLE is four bits.
Broadcast network	A network composed of radio and/or television broadcasting stations. The major broadcasting networks are the four giants: ABC, NBC, CBS, and PBS.
Broadcast videotext	The same as TELETEXT. One form of broadcast videotext uses the so-called *vertical blanking interval* between television picture frames to send videotext information. Most broadcast videotext schemes require the addition of yet another black box to your softspace, and additional monthly charges. For that reason, they are to be avoided in favor of microcomputer-based retrieval methods.
Bug(s)	Error(s) in programs or hardware.
Bulletin Board System (BBS)	A telecommunications service that allows users to place messages to others in a micro's memory, simulating a physical bulletin board. The messages are arranged in a linear string and must be accessed in sequence, one at a time, in the order in which they were added to the system. Some BBS systems allow the creation of different categories of messages (Help Wanted, For Sale, etc.). BBS systems most resemble electronic versions of a newspaper's classified

advertising section. BBSs may also sometimes be referred to as CBBSs—*C* for Community.

Byte A single CHARACTER, composed of eight BITs, four SLICEs, or two NYBBLEs. The byte is commonly used as a unit of measurement for various computer characteristics, such as the amount of memory storage available.

CBMS Computer-Based Message System. A nice generic acronym that can cover bulletin board systems, electronic mail, teletext and videotext systems, and so on. In short, any communications system that uses a computer somewhere in its process. This term is going out of use fast.

CCITT Comité Consultatif Internationale de Télégraphie et Téléphonie. An international consultative committee that sets global communications standards.

CPE Customer-Provided Equipment; Customer-Premises Equipment. This usually refers to the telephones and other equipment you own, connected at your location, and not leased or provided by the telephone company. The phone companies call anything connected at the ends of their wires *terminal equipment*.

CP/M—CP/M-86 Control Program/Microcomputer. A widely used *operating system* for Z-80-based microcomputers. The operating system is the communications traffic director for micro information systems. CP/M-86 is the newest and slickest version for 8086/8088 micros.

Carrier signal The signal on which information can be impressed for transmission. A carrier signal is MODULATED and DEMODULATED according to fixed PROTOCOLS. An example of a carrier signal is the set of tones generated by a MODEM for transmission on the regular phone lines.

Ceefax A TELETEXT system operated by the British Broadcasting Corporation (BBC).

Cellular Radio A form of radio transmission and reception that permits continuous communication over a given region. The region is divided up into geographically equal areas, each of which has a transceiving station. The system detects movement of a mobile unit and automatically switches its transmission to the area into which the mobile unit has been moved. This is the latest form of mobile telephone service.

Censorship The denial or curtailing of the rights of free speech and/or of rights and privileges one expects pertaining to freedom of communications under democratic forms of government.

Centronics A manufacturer that has lent its name to a communications standard or PROTOCOL for PARALLEL TRANSMISSION used primarily between the company's printers and microcomputers. It came about because this company managed to place a lot of its printers in the field, thus helping to spread this de facto STANDARD.

Channel A pathway or connection between two or more points through which signals are transmitted. The pathway may be cable, wire, radio signals, sound signals, light signals, or any combination of these.

Character A single letter, number, or other symbol. See ALPHANUMERIC.

Circuit In telecommunications parlance, the same as a CHANNEL.

Circuit switching In telecommunications, the technique of arranging lines around switching equipment. Incoming wires from users' premises are connected to banks of electromechanical or digital switching systems. The alternative is MESSAGE or PACKET SWITCHING.

Coaxial cable A cable composed of an insulated central conducting wire wrapped in another cylindrical conducting wire. The whole thing is usually wrapped in another insulating layer and an outer protective layer. Coaxial cables are used in some kinds of LANs.

Code Any transformation or representation of data in different forms, according to a prearranged convention. ASCII is a form of digital code. Codes may also be private, so that others may not access the coded information, in which case the message is said to be ENCRYPTed.

Common carrier An organization or business providing communications services to the general public at rates set by approval of the FCC.

Community of consciousness In NETWEAVING, a grouping of individuals and/or organizations sharing common interests and concerns. A SPIN is a community of consciousness located in the same relative geographical area.

Compunications A tortured English neologism signifying the use of computers in communications. Avoid this term. It's not very pretty, hard to say, and doesn't say anything new. See TELEMATICS instead.

Computer conferencing A computer-based analog of a face-to-face conference. Participants send and receive messages through a central or "host" micro that organizes the messages, keeps track of usage, and otherwise facilitates the exchange of information. In most micro conferencing systems, messages and new information can be organized around topics and keywords that make retrieval and keeping up with new information simple and easy.

Conference tree The generic name of any COMPUTER CONFERENCING system that allows the creation of new topics or conferences by branching off old ones, forming a "tree-shaped" database.

Contelligent system The system that results when a human being and an information system are conjoined to perform certain tasks or functions. The machine serves to extend the human's mental powers in specific ways, while the human brings consciousness and purpose to the system.

Control character(s) In micro communications, usually an ASCII character designed to cause things to happen in the device to which the control character is sent. For example, certain control characters sent to a printer will cause a carriage return or a linefeed. Modems and video screens are also manipulated using control characters. Other protocols in addition to ASCII also define control characters as part of the required set.

Copyright As usually applied, the rights to intellectual property as recognized and sanctioned by law. An intangible, incorporeal right granted by statute to the creator of artistic and literary productions, whereby he is invested, for a limited amount of time, with the sole and exclusive privilege of causing copies to be made, publishing them, and selling them. This right has recently been granted also to the creators of software products.

Crossover cable Same as NULL MODEM CABLE.

Cryptology The study of the making and breaking of secret codes. Formal cryptology is a branch of mathematics, involving transformations of information according to certain rules. Decryption is the decoding process, usually done by the party for whom an encrypted message was intended. Encryption, of course, puts the message into code. *Anticryption* is the science of sending clear, unambiguous, easily understood messages. Anticryption is useful for those attempting to contact alien intelligences outside the Earth (exopsychologists) and for those attempting to create a bridge between technology and people (technotranslators).

Cybernetics	The study of the relationships between communication and control in organisms and machines. The principal concept of cybernetics is FEEDBACK.
DES	Data Encryption Standard. A DATA ENCRYPTION scheme developed by the National Bureau of Standards. The DES has been widely criticized as too weak to protect data from decryption by large computers, especially those used by government and military intelligence agencies.
Database	Files organized so as to allow computer users to manipulate and retrieve data.
Data communications	The transfer of digital signals through various electromagnetic channels: telephone lines, satellites, microwaves, fiber optic cable, etc.
Data encryption	The translation of information into a CODE or codes readable only by the intended recipient of the information.
Data set	The classic and now archaic name for MODEM of the acoustical variety.
Debugging/debug	The process of finding and removing BUGS (errors), primarily in computer software. It is also proper to speak of debugging a piece of hardware.
Decrypt/decryption	Decryption is the process of changing an ENCRYPTed signal back into usable (readable, meaningful) form.
Defamation	The taking or harming of someone's reputation. The offense of injuring a person's character, fame, or reputation by false and/or malicious statements.
Delta modulation	A digital communications technique for transmitting voice information on digital channels. It is used in some digital transmission systems.
Dibit	A single signal change or *signal event* representing two BITs each (or one SLICE). There are four possible states—kinds of changes or signal types—needed to transmit each of the four dibits: 00, 01, 10, 11. High-speed MODEMs transmit dibits which are then translated back into bits at the receiving end. See QUADBIT and TRIBIT.
Digital signal	An electrical signal that consists of discrete variations in voltage or duration. In computers and data communications, only two states need to be defined in order to transmit digital information. (Of course, if you have more than two states or changes in signal to work with, all the better.) See BIT and ANALOG SIGNAL.
Direct-connect modem	A MODEM that connects to the telephone lines using a modular plug or wired directly to the outside phone line. It thus transfers electrical signals directly to the phone network without any intermediary protective device. Direct connect modems must be certified by the FCC. Direct-connect modems are more reliable than ACOUSTIC-COUPLED MODEMs.
Distributed network	A network that functions independently of its various NODEs. Here we include the idea of REDUNDANCY of information and command. Crucial information is distributed geographically throughout the network/system.
Docent	In its original form, a museum tour guide. In the emerging telecommunications-based global information SOFTSPACE (information ecology, information envi-

ronment), docents will become cultural guides to databases, interaction and search/research strategies.

Domsat Domestic satellite communications.

Download When two computers are communicating, the computer that is receiving a chunk of data or program is said to be *downloading* the information. If, on the other hand, the computer is sending the data, then it is said to be UPLOADing. "Downloading" and "uploading" are relative terms, but usually you can figure out what is meant by the context in which these terms are used.

Dumb terminal A device for sending and/or receiving data over a telecommunications channel that has no processing power of its own; opposed to a SMART terminal.

Dyad A label referring to a relationship consisting of two individuals; contrast with MONAD, a single individual, or a TRIAD, a three-individual relationship. Much of the theory of games applies to dyadic relationships.

EBCDIC (pronounced "ebb see dick") Extended Binary Coded Decimal Interchange Code. An eight-bit character coding scheme used mainly by IBM. The code allows 256 different characters. In order to communicate with some mainframe computers, it may be necessary to use a TRANSLATION TABLE to convert ASCII to EBCDIC and back again.

EIA Electronic Industries Association. A trade organization composed of manufacturers, designers, and marketers of electronics hardware. This organization is influential in the setting of STANDARDS for electronic equipment.

Echo/echoplex In telecommunications, the transmission of the received character or signal back to the transmitting device in order to verify accurate transmission of the data.

Electronic cottage A term coined by futurist author Alvin Toffler *(The Third Wave)* to denote the growing phenomenon of home-produced work, usually based in some form on the new information technologies.

E-mail Electronic mail. A metaphor comparing the electronic distribution of one-to-one messages using computers to the paper and print and envelope transmission of first-class letters through the post office. About as useful a term, in itself, as CBMS.

Encrypt/encryption Encryption is the process of changing an ordinary message into one that can be read only by first decrypting it. The point is to protect the privacy of the sender and receiver of the message. A type of CODE. Synonyms are *encypher/encyphering* for putting material into code. *Decyphering* is then the process of translating the material back into readable form.

Error detection In digital communications, the process of finding and correcting transmission errors that may have been introduced by NOISE in the channel. There are many levels and forms of error detection and correction. See PARITY.

Even parity A form of PARITY error checking in which the number of 1s in a CHARACTER, when the parity bit is included, is always maintained as an even number.

Exchange A geographical area with specified boundaries established for purposes of administration of communications services within that area. An exchange generally includes one or more central offices and associated switching facilities.

Exhaustivity A measure of how completely the concepts within a document have been indexed.

Expert system   A computer/software/memory system that emulates or duplicates the thought processes of a human expert in a specific area. Examples of expert systems are diagnostic systems in medicine and geological analysis systems in the petroleum and mining industries.

FCC   Federal Communications Commission. A federal agency established by the Communications Act of 1934 to regulate communications. It consists of seven commissioners, four bureaus, and associated administrative personnel.

FSK   Frequency Shift Keying. A form of MODULATION used in low-speed MODEMs in which different frequencies are used to represent the 0s and 1s of digital data.

Fax   Abbreviation for facsimile communication. Fax equipment sends and receives printed information (hardcopy) over the telephone lines. This is "paper-to-paper" information transfer. Some fax services are beginning to allow access by microcomputer.

Feedback   The return of part of the output of a system to its input, in order to control the output and keep it within certain defined limits. An example of feedback is the thermostat on a home heating system, which turns off the furnace when a preset temperature has been reached, and turns it on again when the temperature drops below the set point. Feedback is the prime concept of CYBERNETICS.

Fiber optics   The use of very-small-diameter glass fibers to transmit information via light waves. In fiber optic systems, cables made of bundles of glass fibers are used for linking devices, similar to the way COAXIAL CABLEs are normally employed. There are many advantages to using light as a CARRIER SIGNAL in this manner: low NOISE, greater BANDWIDTH, and lower cost per unit of transmitted signal among them.

Filter   In telecommunications, a filter is designed to remove NOISE from SIGNALs. In information systems, a filter may be used to keep out information not intended for a particular destination or address.

Forth   A high-level programming language. More flexible than BASIC, easier to learn than Pascal. There are over 200 computer languages, and ten times that number if one counts dialects, subsets and supersets, offshoots, and so on. BASIC is ubiquitous, and new versions of BASIC are still being written. Pascal is a popular language used in teaching computer science in universities, and Forth is too powerful to ignore. Now you know why *your* favorite computer language is not listed in this glossary.

Freeware   As used in the BBS community, freeware is public domain software distributed by clubs and individuals at no or low cost to the new user.

Full duplex   In telecommunications, a mode of data communication in which both parties or devices have access to the transmission channel at the same time, thus allowing simultaneous two-way information exchange. Normal voice telephone conversations are usually full duplex. See HALF DUPLEX and ECHOPLEX.

Gaia hypothesis   The Gaia hypothesis—from Gaia, Greek goddess of the Earth—holds that the Earth's atmosphere is a circulatory system and regulating medium maintained by and interacting with all life on the planet. This medium is highly sensitive to changes in its composition. In effect, the hypothesis conceives of the whole planet as a single life form. The theory was first formulated by scientists James Lovelock and Lynn Margulis in 1975.

Games, theory of
A mathematical tool for the analysis of human social relations. Introduced by von Neumann in 1928 and originally applied to decisionmaking strategies in economic behavior, it is now extended to many kinds of interpersonal behaviors. In netweaving, it is well to keep in mind the two basic kinds of games:

1. Zero-sum games: These are situations in which the gain of one player or node and the loss of another always add up to zero. This is the *pure competition* case.

2. Non-zero-sum games: Situations in which the gain of one party is not necessarily the loss of another, and thus gain and loss do not necessarily sum to zero. Such games may be further classified as *pure collaboration* or *mixed motive* games.

Gateway
A NODE that provides a link between two or more NETWORKs. The gateway node translates the PROTOCOLs of one network into the protocols of another, so that the parties at either end can communicate with one another.

Gateway city
A city where GATEWAYs between networks exist for access to connections with other networks or overseas communications services.

Gestalt
Form, pattern, structure, or configuration. For example, in the figure below, each individual dot is an element of a gestalt which we interpret as the Roman letter *F*.

```
  .   .   .   .   .
  .
  .
  .   .   .   .
  .
  .
  .
```

Gray lit/gray literature
Documents and publications that are not formally listed and priced, such as institutional reports, white papers, and privately circulated monographs. Referred to as such by librarians and database managers because such items are difficult to trace. Much of the intelligence at the cutting edge of science and technology exists as gray literature.

Group
Two or more people who interact with one another and who recognize themselves as a distinct social unit.

Group dynamics
The study of human interaction in groups, or *groups as systems* in and of themselves. The simplest form is the DYAD. Small groups are composed of no less than three (TRIAD) nor more than fifteen people.

Guard band
In telecommunications, an unused band of frequencies between two adjacent channels providing a safety margin to prevent adjacent frequencies from interfering with each other.

HDLC
High-Level Data Link Control. An international communications PROTOCOL defined by the International Standards Organization (ISO).

Half duplex
In telecommunications, a mode of data communication in which only one party

or device has access to the transmission channel at any time. This mode of communication requires some kind of "end of transmission" signal so that the channel can then be turned over to the other party.

Handshaking In telecommunications and data communications, the initial exchange of signals between two devices. This exchange helps the two systems set the stage for the transmissions that follow by establishing PROTOCOLs between them. Handshaking can be done at the physical level (hardware handshaking) or the software level.

Hertz (Hz) A unit of frequency named for German physicist Heinrich Hertz. 1Hz = 1 cycle per second (cps).

IEEE-488 An interface standard defined by the Institute of Electrical and Electronics Engineers (IEE). The standard defines a parallel interface, just as the RS-232 standard defines a serial one.

Informatics A term derived from the French term *informatique,* which covers the broad range of information, information handling, and new technologies dealing with these things.

Information The *meaning* that is assigned to what otherwise is mere data. Conventions or PROTOCOLs for the representation and/or interpretation of data enable communication to take place in human-machine and machine-machine systems. In information theory, there is also a mathematical definition of information, enabling information content of messages to be quantified.

Infosphere The organized layer of intelligence-produced information artifacts on planet Earth. Consisting of people, NODES, NETWORKS, INFORMATION, and SOFTSPACES.

Intelligence amplifier Any mechanical or electronic device that augments a human's intellectual or information processing capacity: sextants, pocket calculators, computers in all their forms.

Interface A shared mechanical or electrical boundary through which energy carrying information may selectively pass. A cell's membrane is the interface between the cell's interior and exterior. "Interface" is also used to refer to particular pieces of equipment designed to link together two or more physical systems. It is *très gauche* to use the term "interface" to refer to a business luncheon, unless you live in New York City.

Interrupt/Interrupt-driven In telecommunications, refers to the allocation of CPU resources on the basis of an active request, or *interrupt,* from a peripheral for service from the CPU. In interrupt-driven systems, the microprocessor only interacts with a peripheral when it is called upon to do so. In systems using POLLING for data transfer, the CPU constantly checks the device—say the keyboard or disk drive—to see if service is needed.

Key telephone system Terminal equipment that allows for the selection of multiple lines from one telephone set. Line holding and interconnection are usually featured, with lights to indicate the status of each line. May be part of a Private Branch Exchange (PBX/PABX).

Kinesics Body language or the study of body language. Kinesics plays an important part in face-to-face meetings. However, in some communications situations it is helpful to eliminate kinesics as a factor. Phone conversations and computer conferencing (without video) do not include kinesic information exchange.

LAN	See LOCAL AREA NETWORK.
Leased line/leased circuit	A telecommunications link reserved for use solely by the leasing party. Same as a private line.
Libel	To defame or injure an individual's reputation or character by a published writing. Defamation expressed in print, writing, pictures, or signs.
Local Area Network (LAN)	A network of data processing equipment residing in the same room or building or group of buildings (hence *local* area). Such a network is usually created using its own wiring system, although some LANs can use in-house telephone wiring or the electrical wiring of a building.
Local loop	The telecommunications paraphernalia that connects a user's premises with the local central telephone office. It is called a local loop when the connection is to a long-distance line or service. Otherwise, it's just an access line or local office line.
Long lines	The division of AT&T that provides interstate long-distance communications services.
MUX/statistical MUX	Multiplexer. A statistical MUX is a multiplexer using statistical techniques to combine several data signals for transmission over a single channel, usually a phone link. See MULTIPLEXING TECHNIQUES.
Macro	As a prefix, as in *macro*system, this modifier designates the large-scale and very-large-scale systems. The global weather system is a macrosystem. As a stand-alone noun, "macro" refers to a stored string of characters, usually a set of phone numbers and access codes, that a given communications software package can call up and use automatically when told to do so by the user. The use of macros saves time insofar as long strings of numbers and codes can be reduced to one or two keyboard strokes.
Mainframe	Any permanently installed, physically heavy computer. This term is going out of use as microprocessors become ubiquitous. Many corporations still run and refer to "mainframe" computers, however. The "main frame" refers, historically, to the rack or frame that held the components making up the computer proper, as opposed to peripheral equipment attached to it. The physical size and weight of mainframes helped give rise to the saying "Never trust a computer you can't lift."
Meme	A basic unit of cultural evolution. Examples of memes: tunes, ideas, catch phrases, clothing fashions, ways of making pottery or building arches. Memes are to the evolution of the NOÖSPHERE as genes are to biological evolution.
Message switching	The process of receiving and storing a message until an outgoing circuit and station are available, then retransmitting it to its destination using address information contained in the message itself. No physical switching matrix is used (as in line switching). Protocol conversion and error detection may be a part of the process. There is usually a slight delay in transmission. This kind of switching is used mostly for data, not voice.
Meta-	A prefix meaning "changed in position," "beyond," or "higher." It can refer to a body of knowledge *about* a body of knowledge or field of study, as in "meta-communication" (information about communication) or "metalanguage" (information about languages). It can also refer to the organizing patterns used in ordering a field of study, as in "metalogic."

M.I.N.D.   Mutual Inquiry Network Development. An ongoing transdisciplinary, multilingual, interplanetary, transpersonal applied netweaving research group.

Modem   Modulator-demodulator. A device that changes digital signals into a form that can be transmitted over a telecommunications channel and back again. Modems connect micros with micros, terminals with micros, and a variety of other devices to micros. Most modems convert DIGITAL signals to ANALOG and vice versa, although there are other kinds of modems used in special cases.

Modulate   This is a key concept in all of telecommunications. It is the impressing of information on any electromagnetic signal. The signal itself is usually called the CARRIER SIGNAL. Information is added to carrier signals by using one or more of three basic aspects of a carrier wave: its amplitude, its frequency, or its phase angle.

Multiplexing techniques   In telecommunications and data communications, methods for transmitting many signals over a single channel and reconverting them into separate signals at the receiving end. There are two common methods of multiplexing. *Frequency-division* multiplexing divides the transmitted frequency band into many narrower bands and uses each of the smaller bands as its own channel. *Time-division* multiplexing switches the channel to each of several inputs successively. *Statistical multiplexing* switches each input into the channel on the basis of probable need.

NAPLP   North American Presentation Level Protocol. A telecommunications PROTOCOL allowing both graphics and text to be transmitted on the phone lines. This protocol was developed in Canada, in the Telidon system, and has been adopted for support by AT&T, IBM, and many other telecommunications companies. Its main attraction is in the quality of its graphics.

Natural language(s)   Ordinary spoken or written human languages. Programming or machine languages are really subsets of natural languages designed for the exchange of information between machines. A *natural language system* is an information system in which ordinary language can be used. High-level programming languages such as BASIC or Forth are closer to natural language than machine codes are.

Netweaving   The process of creating new networks using hardware, software, connecting methods, and information for individuals, groups, businesses, pesonal use, etc.

Network   A system of micros and/or micro peripherals connected by telecommunications channels.

Networking   The process of utilizing existing telecommunications networks for one's own purposes.

Neophile   According to exopsychologist Dr. Robert Anton Wilson, all humans exhibit degrees of change-embracing and change-resisting behaviors. People who like new things, in effect who seek and encourage change for themselves and others, are neophiles. On the other hand, individuals who resist change are termed neophobes. Note that neophiles do not necessarily throw out the old and embrace the new for newness' sake, nor are those who resist change always reactionary in their resistance. Some new things ought to be passed by. Much of what is old is definitely on our side.

Neophobe   See NEOPHILE.

Node A point in a network at which a high density of information and information processing equipment is physically located. A receiving/integrating/transmitting center.

Noise In telecommunications, refers to unwanted additions to signals. In information systems, refers to retrieved documents or other information that do not deal with the required subject.

Noösphere The organized layer of intelligence on planet Earth. Gaia's thought patterns. The term was coined by French biologist-priest Teilhard de Chardin.

Noöspherics The study of the macro-effects and system properties of the organized layer of intelligence and information/energy transforms surrounding the whole Earth. The study of planetary consciousness. Information acting on information.

NTIA National Technical Information Service. A government agency concerned with disseminating technical data generated by government-sponsored research and foreign technical information.

Null modem cable A specially wired cable that links two usually identical microcomputers. The cable connects the two systems so that each "believes" it is talking to a modem. A null modem cable is usually connected between the serial ports of the two devices.

Obscene The Supreme Court definition of obscene is "that which, to the average person applying contemporary community standards, the predominant appeal to the matter, taken as a whole, is to prurient interest, i.e., a shameful or morbid interest in nudity, sex or excretion, which goes essentially beyond customary limits of candor in description or representation of such matters and is a matter utterly without redeeming social importance."

Odd parity A form of PARITY error checking in which the number of 1s in a CHARACTER, when the parity bit is included, is always maintained as an odd number.

Office automation The organization of existing computer and network technologies to perform clerical tasks such as document preparation (word processing), filing (data base management), and communications (electronic mail). Office automation is targeted squarely at middle-echelon bureaucrats in an attempt to increase their productivity.

Off-line In telecommunications systems, a device is said to be off-line if it is not immediately accessible by users of the system without human intervention to put it ON-LINE. A component that is off-line can sometimes be operated independently of the system in question.

On-line In telecommunications, a device is said to be on-line if it can be accessed by users of the system without further human intervention. A user is said to be on-line while accessing a remote micro or database.

Packet radio A telecommunications system using microcomputers and small radio transceivers. The message itself is broken up into smaller units or packets. As each packet is sent, the address of the packet is sent along. It is then transmitted over an assigned radio channel, where it can be switched and/or routed so as to make maximum use of the available channel facilities. Advantages of packet radio: many devices can use the same channel without major interference from each other; packets can be sent practically error-free; each node is treated as an equal in the network, no central (host) node is required.

Packet switching	A method of data communications in which the message is split up into smaller units called "packets" and transmitted through the network and then reassembled. Each packet may travel by a different route, depending on channel availability, but it is reassembled at the receiving end, so no one can tell the difference.
Parallel transmission	In data communications, the process of sending all the BITs in a BYTE down the CHANNEL at the same time.
Parity bit	In data communications, a BIT added to the group of seven that define an ASCII character. The value of the bit is specified in advance so that the receiving micro can tell if an error was made during the transmission of any of the seven defining bits. A change in any one of the bits will always be detected. Two or more errors, however, will not necessarily be detected using this simple form of ERROR DETECTION.
Parity checking	An ERROR DETECTION technique in data communications in which the bits in a byte received by a micro are added together to determine whether the sum of the 1-valued bits is even or odd. The parity is specified in advance as being either high (even) or low (odd). The parity bit is added to the character upon transmission by the transmitting micro to make the character conform to the prespecified parity. If the receiving micro comes up with the wrong parity when it makes its own check, then it "knows" an error has been made in the transmission of one of the bits, and it can request retransmission of the character.
PBX/PABX	Private [Automated] Branch Exchange. A telephone switching and connecting center within a company or building.
PCM	Pulse Code Modulation. A method of changing voice information to digital form so that it can be transmitted on digital channels. The signal is sampled periodically and its voltage is converted into a binary number, which is then handled by computers in the communications network.
PIRM	Personal Information Resource Management. Refers to the field of individual information systems design and management, including content assessment.
PSTN	Public Switched Telephone Network. The ordinary everyday voice phone network.
Polling	In LANs, a means of determining whether a particular device is waiting to deliver a message to the network for distribution to another device or series of devices. Each device on the line is "questioned" to determine whether or not it needs service from the network controller. The controller sends a polling signal asking, in effect, "Do you have anything to transmit?" On-line polling systems work in a similar manner. Each user's terminal (or remote micro, as the case may be) is polled by the host computer via the incoming phone line. As more and more users are added to the system, polling takes longer and can become noticeable to an on-line user, experienced as delays between commands or requests sent and the accessed computer's response.
Power	The combination of awareness, social contexts, and resources that results in the ability of a group or individual to influence or control the behavior of others.
Prestel	A TELETEXT service of the British post office, begun in 1974–75. This early version was called VIEWDATA: one-way, mass-market-oriented on-line information service using the standard home television set as the display device.

Printerfacing	The process of connecting a microcomputer to a hardcopy printer. This part of an information system is often the most time-consuming part to install and the most intricate to maintain, since printer devices vary in capabilities and design and are electromechanical as opposed to all-electronic. Mechanical devices, with their moving parts, fail more often than electronic ones.
Privacy, right to	The right to be left alone, free from unwarranted publicity of one's personal affairs.
Private line services	A telephone line and associated equipment provided for the exclusive use of a single subscriber or company, or a specified group of users who do not interfere with one another in the use of the line. Private lines are graded by the telco into the following categories: sub−voice grade telegraph channel; voice grade channel for voice; voice grade channel for data; high-speed facsimile channel; audio program channel; video program channel; and broadband channel, capable of being subdivided for several purposes.
Private network	A network owned, maintained and used by a private individual or company.
Protocols	In telecommunications, the set of conventions that defines the formatting of data. Protocols can be expressed in STANDARDS for hardware and can also be expressed in software. Any two machines must use the same protocols in order to exchange any information at all. See HANDSHAKING and ASCII.
Quadbit	A single signal change or *signal event* representing 4 BITs each (or 1 NYBBLE). There are eight possible states—kinds of changes or signal types—needed to transmit each of the eight quadbits. Some high-speed MODEMs transmit quadbits. See DIBIT and TRIBIT.
Query language	A high-level programming language that resembles natural language closely enough to make on-line searching and retrieval of information from databases easier for nontechnical persons.
RCP/M	Remote CP/M. A form of BBS using the CP/M operating system and designed to be most useful to other CP/M-based information systems.
Redundancy	In information theory, the proportion of information in a message that can be eliminated while still maintaining the essential meaning of a message. In information systems, the provision of a backup system or backup data in case of loss of the original is an example of data redundancy.
Ring back system	There are two different forms of computer-to-computer ring back systems. In the first case, the system making the call connects to the called computer, enters an identifying code, and then hangs up. Based on the identifying code, the second computer "rings back" the first. This is a form of access security, making sure that only a specified location can connect to the called computer. The second kind of ring back system is used when a computer shares a line that is also used for voice, which is the case for many hobbyist BBS systems. Here, the caller dials, lets the phone ring once, and then hangs up. The caller then redials the number, and the micro at the other end answers, having been clued in that the next call is a computer call and should be answered with a carrier signal. Otherwise, a human answers the phone.
RS-232/RS-232C	The designation applied to a STANDARD defined for asynchronous serial data communications between any two devices. The standard was developed jointly

by the Electronic Industries Association (EIA), the Bell System, and computer and modem manufacturers. The definition includes the signal characteristics to be found on each pin, but does not specify any particular connector. The DB25 physical connector has become the conventional one used in most RS-232C cables. Because this "standard" has been haphazardly adopted, there have been and likely will continue to be variations from manufacturer to manufacturer and country to country. See CENTRONICS.

SPIN
See SPECIAL-INTEREST NETWORK.

Serial interface card
A printed circuit board that plugs into a microcomputer. The card changes the parallel internal communications of the micro into a one-bit-at-a-time serial transmission for sending through a MODEM.

Serial Transmission
In telecommunications, a mode of transmission in which the BITs in a BYTE are sent one at a time through a given channel.

Signal(s)
Physical phenomena used to convey data, such as electrical pulses, sounds, light, etc.

Sine waves
Signals of unvarying pitch (frequency) or strength (amplitude). The CARRIER SIGNAL used in MODEMS uses sine waves of specified frequencies (Hz).

Slander
The nearly verbatim legal definition of "slander" is "a false and unprivileged legal publication, orally uttered, and also communication by radio or any other mechanical means which (1) charges any person with a crime; (2) imputes in him or her the present existence of an infectious, contagious, or loathsome disease; (3) tends directly to injure him or her in respect to the office, profession, trade, or business either by imputing general disqualification in those respects which the office or occupation peculiarly requires, or by imputing something with reference with his or her office, profession, trade, or business that has a natural tendency to lessen its profits; (4) imputes to him or her impotence or want of chastity; or (5) which, by natural consequences, causes actual damage."

Slice
Two BITs. See NYBBLE and BYTE.

Smart
An adjective anthropomorphically applied to machines that have some processing capabilities of their own, such as modems or printers or terminals.The "smarts" are usually the result of adding one or more microprocessors to the device. In that sense, certain automobiles can now be called "smart cars."

Softspace
Refers to individual combinations of conceptual and physical spaces. The conceptual spaces consist of information, methods of organization and retrieval, and GESTALTs formed out of individual experiences. The physical spaces house the mechanical and technical paraphernalia of information gathering, organizing, and management.

Software
The instructions that tell a computer what to do. Also may refer to the information contained on such media as videotape or digital disks or stereo records. "Books are software for the mind."

Special-Interest Network (SPIN)
A network extending across a city, state, or other geographical region and linking communities of consciousness and interest throughout the linked area.

Specialized Common Carrier (SCC)
A COMMON CARRIER, usually intercity, authorized by the FCC to provide specialized communications services. SCCs usually provide service to high-density, low-cost intercity private lines, e.g., Southern Pacific Communications' Sprint service or MCI.

Standard(s)	In telecommunications, usually a physical specification that sets forth the PROTOCOLS and conventions to be used in a particular device or in SOFTWARE that is supposed to perform a specific function, such as remote terminal communications. Currently, there is a proliferation of so-called de facto standards that got the status of being "standards" simply because a lot of people used them. While there is some advantage to having standards in the long run, in the short run the lack of standards is probably a good thing, insofar as there is still room for innovation and technical improvement in most areas of telecommunications work. Some "standards" are proprietary to particular companies and should thus be avoided.
Switched network	Refers to all the components required to get a message from point A to point B. The system is assumed to be integrated. A given switched network is usually public and is only part of the entire telecommunications network. Different services, such as telephone and telegraph, use their own switched networks.
Synchronicity	A meaningful coincidence. An event in which two or more previously unrelated events, pieces of information, or other occurrences form a highly charged and meaningful GESTALT in the mind of the perceiver of the events.
SYSOP (pronounced "sis-op")	System operator. A person who oversees the day-to-day operations of a network facility. Sysops may run BBS systems, teleconferencing systems, or large packet switching facilities.
System	In telecommunications, a group of people, machines, and channels organized to accomplish a specific communications purpose. In systems theory, a specific entity with a given boundary and set of relationships within the boundary and to the outside environment.
System, social	Two or more individuals or collective social components sharing relationships and boundaries. Social systems include groups, organizations, societies, and supranational entities.
Systems analysis	Any of a number of techniques employed to analyze existing systems and determining the course of transformation of such systems for purposes of increased creativity, productivity or efficiency. Usually, the transformation will involve the installation or modification of microcomputer-based subsystems.
Systems theory	A body of related definitions and assumptions that treats physical experience ("reality") as an integrated whole, composed of hierarchies of matter and energy organized by information.
Synchronous transmission	A data communications mode in which each string of BITs that make up a CHARACTER (BYTE) is transmitted in accordance with a timing signal—in synchronization. A timing marker signals the beginning of a given character, and the bits are transmitted in strictly timed intervals.
Tariff	The specification of the rates to be charged to users by COMMON CARRIERS. Common carriers are required to file a tariff with various regulating agencies such as the FCC.
Technology assessment	The attempt to study existing technological possibilities and trends with a view toward assessing possible impacts on economies, social systems, and ecologies.
Telco	The telephone company. Local Bell operating systems and/or AT&T.

Telecommuni-cation(s)	Any communication via electrical or electromagnetic media. "Telecommunications" generally refers to radio, television, telegraph, telephone, and digital systems. Current usage in the microcomputer world tends to confine telecommunications to digital signals exchanged between computers, although telecommunications can also embrace almost any transmission of signals, including those on floppy disks and video cassette tape.
Telecommuting	The use of telecommunications and information technology to replace or cut down on daily travel to a work space. The number of people telecommuting to work each day is growing in the U.S.
Telecomputing	An obsolete term referring to the use of computers in combination with a telecommunications channel. See TIME-SHARING.
Teleconferencing	The general term for any conferencing system employing telecommunications links as an integral part of the system. There are four basic types or subdivisions, although these are beginning to merge in various permutations: video, computer, synchronous, and asynchronous.
Teledelivery	The use of telecommunications channels to deliver information products and services.
Telemarketing	The use of the telephone, computers, and video to market products and services. At its most primitive, telemarketing involves sales calls to prospective customers.
Telematics	Information technologies and their applications. Derived from the French *télématique*.
Teleordering	The use of the telephone, micro, and television to order products and services.
Teletext	A form of BROADCAST VIDEOTEXT. Data is sent one "page" or "screen" at a time and serves its users in a one-way fashion. Users "snatch" the information they want from an ongoing stream that repeats itself over and over again. Such a system uses existing transmission media, sandwiching its signals in unused portions of the overall signal. The British Broadcasting System's CEEFAX system works this way.
Telex	A global telegraphic service established by Western Union. A *telex machine* communicates using BAUDOT code at about 50 baud.
Terminal	A device for sending and/or receiving information over a telecommunications channel.
Throughput	The number of units of data (BITS, BYTES, BLOCKS, messages, CHARACTERS, words, calls, whatever) that pass through all or part of a given information system when the system is working to its fullest capacity—sometimes also called its *saturation* point. Throughput is usually expressed as units per given time interval, such as bits per second (bps) or blocks per second or messages per day.
Time-sharing	An archaic term referring to the apparently simultaneous use of a computer by several users from remote terminals. These terminals may also be computers or microcomputers and may be linked to the central computer via the telephone network. See TELECOMPUTING.
Touch Tone	A registered TRADEMARK of the Bell System. Refers to the system of tones used to activate computerized switching stations throughout the telephone network. The tones generated by a Touch Tone telephone set are a subset of the tones used in the network.

Trademark
With reference to registration of trademarks, "trademark" includes "every description of word, letter, device, emblem, stamp, imprint, brand, printed ticket, label, or wrapper, usually affixed by any mechanic or tradesman, to denote any goods to be imported, manufactured, produced, compounded or sold by him, and also any name or names, marks or devices, branded, stamped, engraved, etched, blown, or otherwise attached or produced upon any cask, cake, bottle, vessel, siphon, can, case, or other package, used by any mechanic, manufacturer, druggist, merchant, or tradesman, to hold, contain, or inclose the goods so imported, manufactured, produced, compounded, or sold by him."

Translation table
A table of values used to convert PROTOCOLs from one to another. The table is used in communications software and automatically makes the necessary changes to the outgoing signal so that the micro or computer at the other end can use the information. Translation tables are especially important in converting word processing files for use by automatic phototypesetting equipment.

Transponder
The relay equipment located on a telecommunications satellite that gets the earth signal and retransmits it to other earth stations.

Triad
In SYSTEMS THEORY and GROUP DYNAMICS, a three-element unit studied as a single entity. See DYAD.

Tribit
A single signal change or *signal event* representing three BITs each. Some high speed MODEMs transmit tribits. See QUADBIT and DIBIT.

Trunk
A communications channel between central offices or switching devices. The existing network of trunks offers alternate routing possibilities for any one message. Trunks can also be leased for dedicated purposes.

UART (pronounced "you-art")
Universal Asynchronous Receiver/Transmitter. An acronym designating the microelectronic "chip" that converts PARALLEL signals into SERIAL signals and back again. It is used on a SERIAL INTERFACE CARD.

User-transparent
A system or method is said to be user-transparent if the user is not aware that the system is being employed. The methods by which a phone call is routed through the phone network is user-transparent insofar as you never need to concern yourself with it when dialing a friend. Similarly, software is said to be user-transparent when it works so automatically that you don't need to pay any attention to it, except perhaps to set it up initially.

Value-Added Carrier (VAC)
A COMMON CARRIER that leases telecommunications links from other common carriers and provides them, along with some other service, to an end user. Sometimes also referred to as a VACC—Value-Added Common Carrier—or VAN—Value-Added Network.

VAN
Value-Added Network. See VALUE-ADDED CARRIER.

Vertical blanking interval (VBI)
The VBI is the "space between the frames" in a broadcast television signal. Much of this space in the signal is unused, so that it becomes possible to pack digital information into this unused portion. Special decoder boxes (MODEMs) must then be used to extract and display the information on an ordinary television set. By its nature, information sent this way is one-way, and hence the VBI encoding is used by BROADCAST VIDEOTEXT or TELETEXT systems.

Videotext
The generic term for any electronic or telecomunications-based system that makes information available via television or video display terminals. Broadcast videotext is also TELETEXT, and is usually a "one-way" design: users may only

select from what is transmitted. Interactive videotext is preferable and allows users to choose specific items from the database, add their own items to the database, and to specify retrieval and search strategies to the controlling computer.

Viewdata   Early original name of the British PRESTEL system.

Voice grade line   Telephone lines that connect most of the telephones used for human-to-human conversations are called voice grade. Voice grade lines permit the transmission of sound frequencies between 300 and 3,400 Hz. In order to transmit digital data on voice grade lines, the data must be converted into signals that can be handled by voice grade lines. This is done using modems.

WATS   Wide Area Telecommunications Service. AT&T's bulk rate, public, switched long-distance phone service. WATS billing is based on the number of hours of use rather than per-call.

Index

SUBJECT INDEX

*Note: boldface numbers show pages where main topics begin.

NAME INDEX

TITLES INDEX

**NETWEAVER'S
FEEDBACK FORM**

The following questions are designed to give you an open feedback loop into our softspace in San Francisco. You can correct our course in midstream with the tiniest of signals. Feel free to either tear out the feedback sheet and fold/mutilate it as you will, or photocopy it, or use your own imagination and make up a new form on your own printer. Send it to:

Netweaver Feedback
CommuniTree Group
1150 Bryant Street
San Francisco, CA 94103

1. Are there things we should add to future editions of *The Netweaver's Sourcebook*? Hardware, software, databases, networks, services, books, periodical articles, groups, trends, ideas, other hot stuff?

2. If you supply info in (1), please tell us briefly why you think it ought to be included?

3. Are there items you think should be removed? Errors, dead addresses, lies, calumny?

4. If you answered yes to (3), why?

5. Tell us what you liked or disliked about this sourcebook, please. Suggestions for improvement?

6. Do you want to be on a mailing list that will be traded/sold/given away to companies and individuals of interest to netweavers? If so, name and address here, please.

7. Do you want to be on CommuniTree Group's mailing list only? If so, then check the box.

8. Finally, if you are interested in keeping in touch with other netweavers on a regular basis, either electronically or through a printed newsletter, please check the box.

(see form over)

1. I think you should add the following: _____

_____

_____

2. Because: _____

_____

_____

3. I think the following should be removed: _____

_____

4. Because: _____

_____

5. I liked/disliked the following about this sourcebook: _____

_____

_____

6. Name _____

Address: _____

City: _____ State: _____ ZIP: _____

E-mail or other electronic access. _____

7. [ ] Please DO NOT SELL, TRADE, OR GIVE AWAY MY NAME AND AD-DRESS! IT IS FOR COMMUNITREE'S USE ONLY.

8. [ ] I am interested in your newsletter for netweavers. Please send details.